Multimedia and Communications Technology

About the author:

Through his work with Motorola Semiconductors, the author has been involved in the design and development of microprocessor-based systems since 1982. These designs have included VMEbus systems, microcontrollers, IBM PCs, Apple Macintoshes, and both CISC and RISC-based multiprocessor systems, while using operating systems as varied as MS-DOS, UNIX, Macintosh OS and real time kernels.

An avid user of computer systems, he has had over 60 articles and papers published in the electronics press, as well as several books.

Multimedia and Communications Technology

Second edition

Steve Heath

Focal Press
Taylor & Francis Group

NEW YORK AND LONDON

First published 1996
Second edition 1999

This edition published 2015 by Focal Press
70 Blanchard Road, Suite 402, Burlington, MA 01803

and by Focal Press
2 Park Square, Milton Park, Abingdon, Oxon OX14 4RN

Focal Press is an imprint of the Taylor & Francis Group, an informa business

British Library Cataloguing in Publication Data
A catalogue record for this book is available from the British Library

ISBN: 978-0-240-51529-8 (pbk)

Typeset by *Steve Heath*

Contents

Preface

For many people, the definition of multimedia is one of a PC playing games or searching an encyclopedia from a CD-ROM. While this is an application for multimedia, it is only a limited view. Multimedia technology is bringing together several existing products such as the PC, telephone and television and combining them in such a way that the borders and definitions that make a television recognisable as such will rapidly disappear. To do this requires a fundamental change in the way the content and information are encoded, communicated and displayed. It is this revolution with developments in PC technology, video and audio compression, telecommunications and many other disciplines that this book addresses.

This second edition is organised as a set of tutorials, starting with the fundamental techniques of digital audio and video and moving through to the compression techniques that are used such as JPEG and MPEG. The delivery systems for multimedia are then covered, starting with the CD and then moving through telephones, local area networks and moving into the new technologies of ATM and ADSL. The final chapters describe how these technologies are brought together in some key applications. The first of these chapters describes video conferencing and the techniques behind the latest H.320 based video phones and conference systems. Digital video broadcasting goes through the techniques of delivering MPEG compressed digital television through satellite, cable and the telephone network and describes the new services that can be expected, such as video on demand and interactive television. The final chapter goes through the multimedia PC in detail with explanations of how the new interfaces such as TAPI and TSPI are bringing PCs and the telephone together. The material has been updated to include information on MPEG 4 with its ability to synthesise both audio and video objects, the new variations on ADSL, the H.323 video conferencing standards, DVD and other new CD-ROM technologies and also includes tutorials on fractal and wavelet compression.

The material in this book is derived from my participation within the Motorola-BT joint development of the Qorus™ video conferencing system over the last two and half years. As can be expected with any work that involves a wide spread of technologies, many individuals in the group have helped me with insight and explanations. I would like to express my thanks to the following people: Pat McAndrew, Colum O'Neill, Dan Fox, Roger Noble, Mike Phillips from Motorola's Aylesbury office; Chitti, Bhaskar and the MIEL team in Bangalore, India; Alistair Buttar and Bernard Dugerdil in Geneva, Jean-Pierre Messa in Paris; and finally, Karl Hebaum and the Kaploweasel in Austin, Texas.

Finally, I would like to thank Mike Cash and Margaret Riley at Butterworth-Heinemann for their help, encouragement and support.

Special thanks must again go to Sue Carter for yet more editing, intelligent criticism and flavoured coffee when I needed it. Without her help and support, this book would have not been possible.

Steve Heath

Acknowledgements

By the nature of this book, many hardware and software products are identified by their tradenames. In these cases, these designations are claimed as legally protected trademarks by the companies that make these products. It is not the author's nor the publisher's intention to use these names generically, and the reader is cautioned to investigate a trademark before using it as a generic term, rather than a reference to a specific product to which it is attached.

The following trademarks mentioned within the text are acknowledged:

MC68000 and Qorus are all trademarks of Motorola Inc.

PowerPC is a trademark of IBM.

iAPX8086, iAPX80286, iAPX80386, iAPX80486 and Pentium are trademarks of Intel Corporation.

PostScript is a trademark of Adobe Systems Inc.

PowerMAC, Apple, AppleTalk, Macintosh, LaserWriter are all trademarks of Apple Computer Inc.

Many of the techniques within this book can destroy data and such techniques must be used with extreme caution. Again, neither author nor publisher assume any responsibility or liability for their use or any results.

While the information contained in this book has been carefully checked for accuracy, the author assumes no responsibility or liability for its use, or any infringement of patents or other rights of third parties which would result.

As technical characteristics are subject to rapid change, the data contained are presented for guidance and education only. For exact detail, consult the relevant standard or manufacturers' data and specification.

1   What is multimedia?

If there is a term or phrase that has appeared in more diverse publications than any other over the last few years, it must be multimedia. The number of definitions for it are as numerous as the number of companies that are working on it. If this is the case, what is multimedia?

In essence, it is the use or presentation of data in two or more forms. The combination of audio and video in the humble television was probably the first multimedia application, but it has been the advent of the PC, with its ability to manipulate data from different sources and offer this directly to the consumer or subscriber, that has sparked the current interest.

As a result, multimedia for many people conjures up the image of a PC with a SoundBlaster card playing interactive games or searching through an encyclopaedia with all the information supplied on a CD-ROM. Whilst this is undoubtedly a valid multimedia application, it is only part of the story. This definition is pragmatic and based on reality; other prophecies for the industry have ranged from the televisions becoming PCs, and vice versa, and that one day, everyone will have a combined televisual-PC-phone-fax, capable of doing everything that you could possibly want and more. Oh and yes, you will never buy software any more, as you will be able to access it from your televisual-PC-phone-fax from the network and only pay for what you need!

The reality is somewhere between the extremes. Undoubtedly, with the ever-improving ability of the PC to provide TV quality audio and video, the television and PC are becoming very close. Add the ability to provide graphical overlays and the difference is very small indeed. With cable TV companies providing telephone connections and the increasing combination of PC with a modem to access the Internet and thus provide an intelligent telephone, the forecasts for the universal widget are a logical progression.

There are some problems involved: the one piece of electronic/electrical equipment that everyone knows how to use and which has a similar interface throughout the world is the telephone. The piece of consumer equipment that causes the most technophobia — after the video recorder remote control — is the PC. The telephone network has evolved over a hundred years and has, through its standardisation procedures, achieved a level of interoperability taken for granted. The PC industry is only twenty years old and is littered with de facto standards that have come and gone within a twinkling of an eye. As soon as something becomes a 'standard' within the PC industry, it is usually a sure sign that it is almost obsolete.

Today, most PCs have high data bandwidth available from a local hard disk or CD-ROM, or from a local area network. This bandwidth is cheap and readily available (for example, disk drive prices for a 1 Gbyte drive have dropped by a factor of 10 in about 2 years) but it can rely on someone buying a CD-ROM containing the material i.e. a slow delivery process.

For faster delivery to the PC, the telephone network that forms a wide area network (WAN) is an ideal candidate — except that it cannot yet deliver anywhere near the bandwidth needed at a low enough cost to replace a CD-ROM or even a satellite connection. However, the telephone network is already available to many potential customers and switching customers to using the telephone link instead of other media, such as cable, would increase the revenue to the telephone company (telco) from the customer. However, this will only happen if the tariffs are competitive with satellite or cable television. The economics for the telcos are such that the cost of replacing the existing network is prohibitive unless network revenues are dramatically increased to pay back the investment. All in all, the whole situation is in balance, with many people juggling many balls in the air. The consumer electronics, software, PC, telephone, television and film companies are all poised at a point where the traditional boundaries will disappear. The problem is trying to work out when this is going to happen and how.

The one thing that is clear is that the days of analogue audio and video are numbered. In the same way that the audio CD destroyed the demand for vinyl records, the introduction of digital audio and video programs, with their associated data to provide sub-titles and so on, will kill the market for analogue systems. The clock is already ticking for analogue broadcasts within the UK, with the decision to switch the currently allocated bandwidth for analogue transmissions to digital broadcasts within the next 15 to 20 years. This decision will allow more channels to fit into the same spectrum — thus increasing the choice available to the consumer.

Multimedia requirements

For a digital multimedia revolution to actually happen, several things have to be in place:

- Demand from the consumer.
  This is often overlooked but, unless there is an advantage for the consumer, multimedia technology will not get the acceptance it needs to satisfy the commercial side of the technology. There is currently a debate concerning what the market actually wants as opposed to the industry's views. Does the consumer want 500 channels or should the focus be on quality rather than quantity? If the consumer is given 500 channels, what tools are going to be provided to help the viewer manage the vast amounts of program information and decide which programs to watch?
- Compression techniques to make transmission viable or reduce the amount of bandwidth needed.

Digital audio and video data consumes a large amount of bandwidth and, unless compressed significantly, it is impractical to deliver in its original format. Key parts of multimedia technology are the compression techniques for audio and video data. This indirectly raises another requirement — to provide the processing power to decompress/compress the data at the user end.

- Processing power to handle the compression/decompression. This can take the form of special hardware or software running on a very fast generic processor. With video compression and reduced bandwidth, hardware support is essential and, although software can be used, the performance is not that good. However, processing power is getting faster and cheaper with time and this is thus an area where a convergence between the two approaches will probably happen.

- Standards.
This is an essential part of the equation. Without the appropriate standards to guarantee interoperability, and an unsegmented market to ensure that the financial returns and incentives are there, the potential for multmedia systems will never be realised. Fortunately, this has been acknowledged and many industry and standards led consortia are addressing these issues.

- Back channels to provide an interactive loop.
The provision of a back channel to allow the consumer to feedback to the service provider is essential to support interactive television and also to support the billing services that will be necessary to allow the consumer to be charged for the services received. The back channel in most multmedia applications does not require a large bandwidth. The one exception to this is video conferencing, where the bandwidth is usually symmetric.

- Bandwidth.
This is probably the most critical area for a multimedia system. Without sufficient bandwidth, multimedia applications are simply not viable. The challenge is not simply in providing X Mbps of bandwidth in the home: the bandwidth has to have the right characteristics to ensure that quality of service is not degraded. The quality required depends not only on the application but also on the data type being transferred. For interactive audio and video, this means guaranteed bandwidth when required. For one-way video, the requirement is less demanding and buffered transmission is adequate.

- Internal distribution.
If the PC, telephone and television are going to become a single integrated unit, each consumer will need an efficient method of distributing this data throughout the household; this, in itself, is a major undertaking. Whilst there have been several at-

tempts at establishing a 'house of the future' standard that provides this routing, it has not yet gone beyond the exhibition show house. To be really successful, each house will require the equivalent of an internal computer network. The cost of installation is immense, even in a new house, let alone retrofitting into an old one. This is due to the costs of cable and associated electronics, neither of which is currently in volume production. It is not clear today how the consumer views this development or requirement, but it may fall into the category of technically possible but not accepted by the consumer.

Multimedia scorecard

Given the scenario of developments and devices, how are the current multimedia systems poised to take part in the revolution?

The PC

Over the years, the PC has become multimedia-enabled with the inclusion of digital audio through the use of sound cards, digital video (through either software or hardware) and access to high bandwidth devices like CD-ROM drives. It has all the interactive support needed and the ability to communicate with others over LANs and WANs. As a self-contained multimedia system, it has much potential. As part of a wider multimedia network, there are limitations on what the device can do. These arrangements are fine for exchanging data, such as documents and files, but although the bandwidth appears to be sufficient, a PC's characteristics, especially on a heavily congested network with many other PCs connected, are not good for multimedia.

The PC is evolving into several variants: the advent of computer-aided telephony, where the PC takes over the functionality of the telephone, is becoming more and more prevalent and producing new products and services. The recent introduction of fax based databases, where data can be faxed from a server to a fax machine is a good example of this alliance. The data is digitally stored on a PC that acts as a server. The commands, sent as tones from a normal telephone keypad, are received by the PC which processes them and sends the appropriate data sheets as fax data.

The audio-visual PC, where movies can be digitally edited, is another variant. This, combined with other techniques, such as PhotoCD and video compression, has reduced the cost of producing digital programs and thus helped provide the material to fill the 500 or so channels that could be on offer. The PC used for producing and utilising interactive electronic copies of books is yet another different aspect of the technology.

The prospect for the PC is that of the universal multimedia system that can, with different software and hardware, support any type of multimedia system needed. This does not mean to say that it will replace the television or telephone (this is unlikely, in my opinion, although some would disagree) but it will be capable of being a telephone or television and thus integrating the functionality into its existing and evolving capabilities.

The television

The television has already undergone a radical change with the delivery of programs. The introduction and availability of satellite, cable and radio broadcasts have increased the scope of available program material.

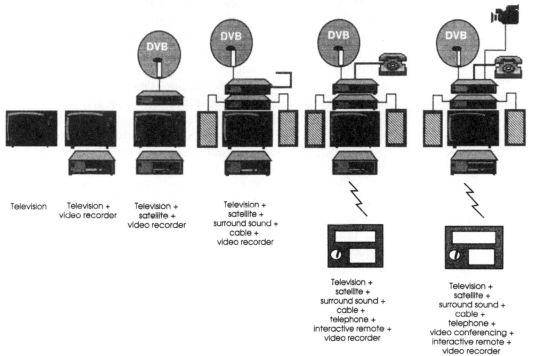

Television   Television + video recorder   Television + satellite + video recorder   Television + satellite + surround sound + cable + video recorder

Television + satellite + surround sound + cable + telephone + interactive remote + video recorder   Television + satellite + surround sound + cable + telephone + video conferencing + interactive remote + video recorder

Possible evolution of the television

The advent of interactive CD (CD-i) based systems that can use a television to display and play back movies, photographs, databases and games has provided an insight into the future of television programs. The combination of audio and video, along with graphics so that data about characters or sports statistics can be shown, is defining how future programs will look. The problem is that of bandwidth. With CD-i, the bandwidth and interaction are local. With most existing television systems, the data transmission is one way and any communication from the consumer is performed via a different route. These back channels typically use telephone connections. The television is likely to evolve into its two essential components: the audio-visual monitor that displays and plays back the picture and soundtrack, and the source of the programme material. This material may come from a multitude of sources in the future, including other multimedia devices within the home, as well as cable, satellite and radio broadcasts.

The telephone

The telephone is essentially a communications device — a method of transferring information from one point to another. It provides a link into the biggest network in the world and, as such, can provide the infrastructure to distribute multimedia information. The problem is currently of bandwidth. The amount of bandwidth needed to send analogue speech is very small and is simply incapable of supporting the bandwidth requirements needed by data and multimedia. This was the situation that the switch to digital telephone services was to address — but even these improvements are not enough. As a result, a lot of work is currently being done to address and resolve these issues. With sufficient bandwidth delivered to the consumer, the telephone takes on a multi-faceted role: it can continue to perform the role of a communications device but can now link PCs together to create widely distributed LANs. It can provide audio visual data and thus an alternative delivery mechanism to cable and satellite. It can use the bandwidth to upgrade the communications from audio to include video and data and it has the benefit of having an existing infrastructure in place.

The picture for multimedia developments is both exciting and daunting: exciting because of all the potential developments and daunting because of the scale of the revolution needed — both technically and socially. The key to making sense of multimedia in all its forms is understanding the technologies involved and their interaction. The first step is to understand how digital audio and video work.

2 Digital audio

With the advent of audio CDs and the proliferation of SoundBlaster cards for personal computers, the techniques behind digital audio have come to the forefront. Although simple in concept, the digitisation of analogue signals, like audio, presents its own set of problems to be overcome.

Digital audio has several advantages over analogue audio:

- Digital recordings do not degrade with re-recording. Each copy is an exact reproduction of the original. At the time of recording, the copy can be compared to the original and any errors detected and sunsequently corrected. This does require access to the original recording however. With a digital recording simply being a stream of numbers, it is possible to protect the recording's integrity using error coding. Virtually all digital recordings use various forms of this technique which will allow the recording to be checked and corrected without access to the original. The result is a digital recording that is an exact replica of the original source whenever it is replayed.

- The recording performance is independent of the recording medium. Digital audio bit streams retrieved from a CD or hard disk will sound the same every time they are played back because the bit streams are exact copies. This is not the case with analogue recordings, where the characteristics of the recording medium can influence the reproduced sound and will degrade with time. This should not be interpreted as meaning that a CD will sound the same on every player — it won't! It means that when the CD is played on the same system for a 1000th time, it will sound the same as when it was played the first time. With a cassette tape, the quality will have changed due to degradation.

- Digital audio is easy to process because the signal processing can be performed by mathematical algorithms. Different effects such as phasing, reverb, echo, and so on, can be achieved by mathematical manipulation of the audio signals in their digital format. Better performance can be obtained in terms of filter roll-off characteristics, dynamic range and signal to noise ratios.

Like most things in life, nothing is for free. Digital audio is not perfect despite these advantages and has several disadvantages of its own compared to analogue audio:

- It requires two conversion stages: one to convert the analogue signals into a digital format and a second to convert the digital signals to analogue.

- These conversion processes can introduce their own types of distortion and defects.

- The digital data requires a far higher density storage than its analogue equivalent. An analogue recording system, such as a cassette recorder, cannot store sufficient digital data to reproduce even equivalent quality audio — let alone the improved quality achievable with digital techniques.
- Whilst effects are simple to achieve using algorithms, a very fast processor is required, which can be expensive compared with an analogue equivalent — albeit with far less performance and flexibility.

Analogue to digital conversion techniques

The basic principle behind analogue to digital conversion is simple: the analogue signal is sampled at a regular interval and each sample is divided or quantised by a given value to determine the number of given values that approximate to the analogue value. This number is the digital equivalent of the analogue signal. An example of this process is as follows: consider an analogue signal that can vary between 0 and 1 volt. The signal level is sampled and gives a value of 0.754 volts. This voltage is then converted to a digital value. The digital value represents the analogue voltage as a set of values from 0 to 9. If the analogue value is greater than 0 but less than 0.1 volt, the digital value is 0. If it is greater than 0.1 volts but less than 0.2 volts, it is assigned a digital value of 1 and so on.

When the digital values are converted back to an analogue signal, the resulting analogue value may actually be anywhere between the upper and lower value used during the comparison. The digital value will give no indication of the exact value, only the range of values that it fell in. As a result, an arbitary value will be used and the resulting analogue signal can have errors in it.

The conversion process effectively takes an infinitely variable analogue signal and converts it a digital signal which has a fixed and finite set of values. The larger the range of digital values that are used, the more accurate the digital representation will be as it will have less or smaller errors because each digital value will represent a smaller set of analogue values. In practice, analogue signals are limited by the resolution of the various components in the system. For example, you can change the amplitude of a signal by 0.001 volt but this change may not be audible to the human ear. This can be exploited with digital audio by essentially using a high enough digital resolution that the errors become very small and not discernible.

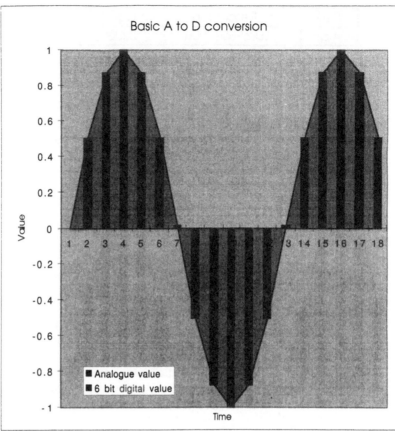

Basic A to D conversion

The result of performing an analogue signal to digital conversion is shown graphically. The grey curve represents an analogue signal which, in this case, is a sine wave. For each cycle of the sine wave, thirteen digital samples are taken which encode the digital representation of the signal.

Quantisation errors

Careful examination of the combination chart reveals that all is not well. Note the samples at time points 7 and 13. These should be zero — however the conversion process does not convert them to zero but to a slightly higher value. The other points show similar errors; this is the first type of error that the conversion process can cause. These errors were mentioned earlier and are known as quantisation errors and are caused by the fact that the digital representation is step based and consists of a selection from 1 of a fixed number of values. The analogue signal has an infinite range of values and the difference between the digital value the conversion process has selected and the analogue value is the quantisation error.

Size	Resolution	Storage (1 s)	Storage (60s)	Storage (300s)
4 bit	0.0625000000	22050	1323000	6615000
6 bit	0.0156250000	33075	1984500	9922500
8 bit	0.0039062500	44100	2646000	13230000
10 bit	0.0009765625	55125	3307500	16537500
12 bit	0.0002441406	66150	3969000	19845000
16 bit	0.0000152588	88200	5292000	26460000
32 bit	0.0000000002	176400	10584000	52920000

Resolution assumes an analogue value range of 0 to 1

Storage requirements are in bytes and a 44.1 kHz sample rate

Digital bit size, resolution and
storage

The size of the quantisation error is dependent on the number of bits used to represent the analogue value. The table shows the resolution that can be achieved for various digital sizes. As the table depicts, the larger the digital representation, the finer the analogue resolution and therefore the smaller the quantisation error and resultant distortion. However, the table also shows the increase which also occurs with the amount of storage needed. It assumes a sample rate of 44.1 kHz, which is the same rate as used with an audio CD. To store 5 minutes of 16 bit audio would take about 26 Mbytes of storage. For a stereo signal, it would be twice this value.

The size of the digital representation is denoted by the number of bits needed to represent it. Digital audio with four bits shows the worst errors while that for 32 bit shows the least errors.

Sample rates and size

So far, much of the discussion has been on the sample size. Another important parameter is the sampling rate. The sampling rate is the number of samples that are taken in a time period, usually one second, and is normally measured in Hertz, in the same way that frequencies are measured. This determines several aspects of the conversion process:

- It determines the speed of the conversion device itself. All converters require a certain amount of time to perform the conversion and this conversion time determines the maximum rate at which samples can be taken. Needless to say, the fast converters tend to be the more expensive ones.
- The sample rate determines the maximum frequency that can be converted. This is explained later in the section on Nyquist's theorem.
- Sampling must be performed on a regular basis with exactly the same time period between samples. This is important to remove conversion errors due to irregular sampling.

Irregular sampling errors

The sampling method can also introduce errors into the digital signal. The conversion should be done on a regular periodic basis so that the further errors are not introduced. Consider a rising wave form such as the beginning of a sine wave.

• If the sample is taken early, the value converted will be less than it should be.

If all or the majority of the samples are taken early, the curve is reproduced with a similar general shape but with a lower amplitude. One important fact is that the sampled curve will not reflect the peak amplitudes that the original curve will have.

• If the sample is taken late, the value will be higher than expected.

• If there is a random timing error — often called jitter — then the resulting curve is badly distorted.

For falling waveforms, the opposite effect is encountered.

Quantisation errors can now compound any of these errors. If the value is taken early and is less than it should be, its converted value could still be accurate. The reason is that although the analogue value being converted is smaller, it is not small enough to cause the digital value to be smaller. The end result is a correctly converted value. Similarly, if the sample is taken later, this may not be big enough to cause the digital value to be larger. The result of these sample timing errors is thus masked by the quantisation process. The conversion process will appear to have worked correctly and accurately. This does not happen all the time. If the correct value is very close to the point where the digital value switches to the next value, even a small jitter may be enough to cause the digital value to be changed and be incorrect. The impact of jitter errors is thus heavily dependent on the analogue waveform itself, the sampling frequency, the digital resolution and the amount of jitter. In some cases, no error may be introduced but in others, this may not be the case. It has even been known for a system to correctly convert at 8 bit resolution but introduce jitter errors when the resolution is increased to 12 or 16 bits. It is important therefore to have a regular and accurate sampling rate, especially when using a high resolution converter. If this is not the case, the end result is not an accurate and faithful reproduction or conversion of the original analog signal.

Other sample rate errors can be introduced if there is a delay in getting the samples. If the delay is constant, the correct characteristics for the curve are obtained but out of phase. It is interesting that there will always be a phase error due to the conversion time taken by the converter. The conversion time will delay the digital output and therefore, introduces the phase error — but this is usually very small and can typically be ignored.

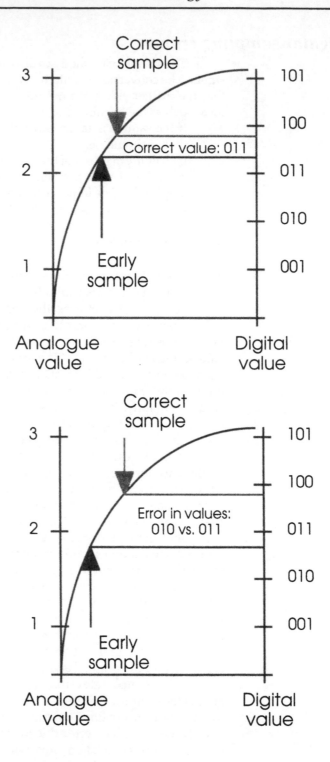

The example on the top shows a case where the early sample
does not change the digital output while the one on the
bottomdoes show how an error can be introduced.

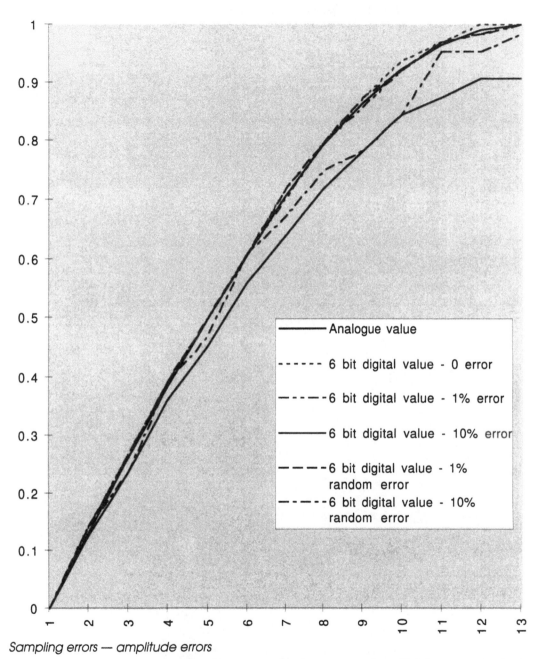

Legend:
- Analogue value
- 6 bit digital value - 0 error
- 6 bit digital value - 1% error
- 6 bit digital value - 10% error
- 6 bit digital value - 1% random error
- 6 bit digital value - 10% random error

Sampling errors — amplitude errors

The phase error shown assumes that all delays are consistent. If this is not the case, different curves can be obtained as shown in the next chart. Here the samples have been taken at random and at slightly delayed intervals. Both return a similar curve to that of the original value — but still with significant errors.

In summary, it is important that samples are taken on a regular basis with consistent intervals between them. Failure to observe these design conditions will introduce errors. For this reason, many micro-processor based implementations use a timer and interrupt service

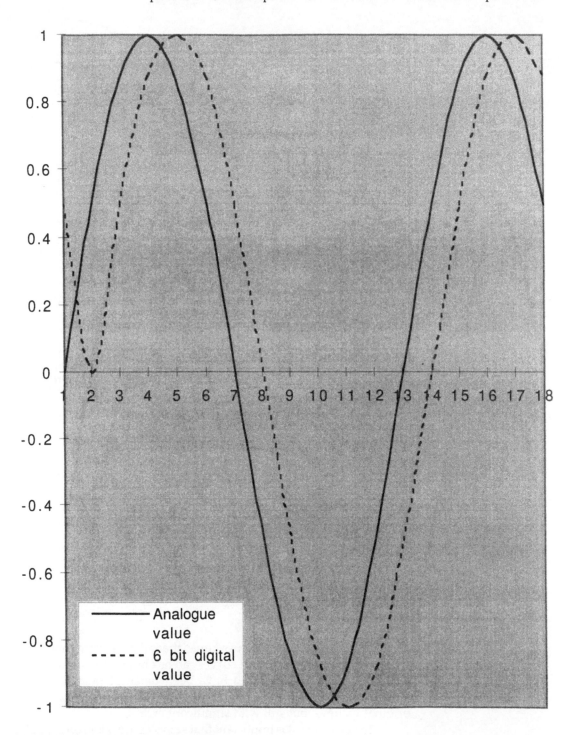

Phase error due to delayed
sample rate.

routine mechanism to gather samples. For a simple system, the timer is set up to generate an interrupt to the processor at the sampling rate frequency. Every time the interrupt occurs, the interrupt service routine reads the last value for the converter and instructs it to start a new conversion before returning to normal execution. The interrupt routine instructions always take the same amount of time and therefore sampling integrity is maintained. This does assume that the time taken from generating the interrupt to servicing it is also consistent. With many of today's RISC based microcontrollers, this is either the case or the timing variation is very small compared to the sampling rate i.e. the jitter error is small. For slower processors that can have instructions that vary in the time to execute, this can introduce unwanted errors.

For more accurate systems, the organisation can be changed. The timer directly controls the converter to start the next conversion and also generates a processor interrupt. The converter will buffer the previous result. The processor will now read this buffered result from the converter. In this way the buffer will remove any jitter. This method does assume that the converter has a buffer and that the processor interrupt routine timing can read the buffer and clear it before the next conversion has completed.

Nyquist's theorem

The sample rate also must be chosen carefully when considering the maximum frequency of the analogue signal being converted. Nyquist's theorem states that the *minimum* sampling rate frequency should be twice the maximum frequency of the analogue signal. A 4 kHz analogue signal would need to be sampled at twice that frequency to convert it digitally. For example, a hi-fi audio signal with a frequency range of 20 Hz to 20 kHz would need a minimum sampling rate of 40 kHz.

Higher frequency sampling introduces a frequency component which is normally filtered out using an analogue filter.

Codecs

So far the discussion has been based on analogue to digital (A to D) and digital to analogue (D to A) converters. These are the names used for generic converters. Where both A to D and D to A conversion is supported, they can also be called codecs. This name is derived from **coder-dec**oder and is usually coupled with the algorithm that is used to perform the coding. Generic A to D conversion is only one form of coding; many others are used within the industry where the analogue signal is converted to the digital domain and then encoded using a different technique. Such codecs are often prefixed by the algorithm used for the encoding.

Linear

A linear codec is one that is the same as the standard A to D and D to A converters so far described, i.e. the relationship between the analogue input signal and the digital representation is linear. The quantisation step is the same throughout the range and thus the increase in the analogue value necessary to increment the digital value by one is the same, irrespective of the analogue or digital values. Linear codecs are frequently used for digital audio.

A-law and μ-law

For telecommunications applications with a limited bandwidth of 300 to 3100 Hz, logarithmic codecs are used to help improve quality. These codecs, which provide an 8 bit sample at 8 kHz, are used in telephones and related equipment. Two types are in common use: the a-law codec in the UK and the μ-law codec in the US. By using a logarithmic curve for the quantisation, where the analogue increase to increment the digital value varies depending on the size of the analogue signal, more digital bits can be allocated to the more important parts of the analogue signal and thus improve their resolution. The less important areas are given less bits and, despite having coarser resolution, the quality reduction is not really noticeable because of the small part they contribute to the signal.

Conversion between a linear digital signal and a-law / μ-law or between an a-law and μ-law signal is easily performed using a lookup table.

PCM

The linear codecs that have been so far described are also known as PCM codecs — pulse code modulation. This comes from the technique used to reconstitute the analogue signal by supplying a series of pulses whose amplitude is determined by the digital value. This term is frequently used within the telecommunication industry.

There are alternative ways of encoding using PCM which can reduce the amount of data needed or improve the resolution and accuracy.

DPCM

Differential pulse coded modulation (DPCM) is similar to PCM, except that the value encoded is the difference between the current sample and the previous sample. This can improve the accuracy and resolution by having a 16 bit digital dynamic range without having to encode 16 bit samples. It works by increasing the dynamic range and defining the differential dynamic range as a partial value of it. By encoding the difference, the smaller digital value is not exceeded but the overall value can be far greater. There is one proviso: the change in the analogue value from one sample to another must be less than the differential range and this determines the maximum slope of any waveform that is encoded. If the range is exceeded, errors are introduced.

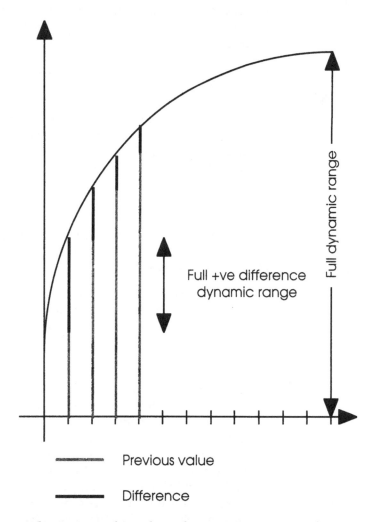

DPCM encoding

The diagram shows how this encoding works. The analogue value is sampled and the previous value subtracted. The result is then encoded using the required sample size and allowing for a plus and minus value. With an eight bit sample size, one bit is used as a sign bit and the remaining seven bits are used to encode data. This allows the previous value to be used as a reference, even if the next value is smaller. The eight bits are then stored or incorporated into a bitstream as with a PCM conversion.

To decode the data, the reverse operation is performed. The signed sample is added to the previous value, giving the correct digital value for decoding. In both the decode and encode process, values which are far larger than the eight bit sample are stored. This type of encoding is easily performed with a microprocessor with 8 bit data and a 16 bit or larger accumulator.

A to D and D to A converters do not have to cope with the full resolution and can simply be eight bit decoders. These can be used providing analogue subtractors and adders are used in conjunction with them. The subtractor is used to reduce the analogue input value before inputting to the small A to D converter. The adder is used to create the final analogue output from the previous analogue value and the output from the D to A converter.

ADPCM

Adaptive differential pulse code modulation (ADPCM) is a variation on the previous technique and frequently used in telecommunications. The difference is encoded as before but instead of using all the bits to encode the difference, some bits are used to encode the quantisation value that was used to encode the data. This means that the resolution of the difference can be adjusted — adapted — as needed and, by using non-linear quantisation values, better resolution can be achieved and a larger dynamic range supported.

Compression techniques

Digital audio requires a lot of storage space — especially when using high sample rates and large sample sizes. To reduce this burden, compression techniques are frequently used to reduce the amount of data needed for encoding.

Byte sized sampling

One technique is to use byte sized sampling. This is not strictly a compression technique, although it can save on processing the data to align it on byte boundaries. Most processors are byte orientated in that they can handle 8 and 16 bit words quite easily and are most efficient at moving data when using data that aligns with byte boundaries. Unfortunately, 8 bit samples are not that high resolution and therefore higher sampling rates are used such as 10, 11, 12 or even 14 bits. Unfortunately, these sample sizes do not fit into byte boundaries and bits are either left unused (which increases the amount of storage required), or data is packed together so that the samples straddle byte boundaries. When this happens, the processor must unpack the data which may require multiple data accesses for some samples that straddle a boundary. All in all, this is not very efficient and thus some systems will only support byte aligned sample sizes such as 8 or 16 bits to reduce the amount of work needed to move the data about and to prevent there being any unused bits.

RLL and Huffman encoding

A digitised audio stream is like any other digital data and can be compressed using any technique used for compressing other digital data, such as graphics or text.

RLL (run line length) and Huffman encoding look for repeated values within a data stream and replace them with a smaller code and thus reduce the amount of storage needed to represent the original data. To extract the original information, the compressed stream is analysed and the codes replaced by the original data. Unfortunately, the efficiency of such algorithms is very much dependent on the contents and sounds with a lot of silence — and therefore consecutive samples with the same value — will compress far better than sounds with little or no silence.

Differential encoding

DPCM and ADPCM techniques can be used to compress a PCM stream, providing its characteristics are suitable i.e. its slope (and therefore change from one sample value to another) is within the range supported by the differential encoding.

There are several ways that these techniques can be used to reduce the effective sample size: if the delta between samples is only a four bit value, then it can be encoded using DPCM where the difference value is only four bits, thus achieving a 2:1 compression ratio. Alternatively, the quantisation for the difference encoding can be changed so that it is a multiple of that normally used: in this way a difference value that needs to be encoded is effectively divided by the multiple and therefore needs less bits to encode it. An eight bit value whose quantisation value is doubled would only need seven bits to encode it. This method can be a little 'lossy' but it can compress the bitstream considerably.

GSM full rate and half rate encoding

The audio compression algorithms used in digital mobile telephones within the GSM network take a different approach to compressing data. They achieve compression by converting speech into a series of coefficients which are fed into a model of the human vocal tract. These coefficients are then used by the receiver within the model to synthesise the speech, complete with the original voice register and intonation. As a result, a normal telephone speech channel can be compressed from its 64 kbits per second requirement for PCM encoding down to 16 kbits per second for the full rate versions and 8 kbits per second for the half rate implementation.

Psycho-acoustic model

This is also a technique that mimics the human aural system. More exactly, it uses the characteristics of the ear to work out where data can be lost so that it is not noticed. This is the basis of the MPEG1 and 2 audio compression algorithms, which are explained in more detail in Chapter 5.

Audio synthesis

This was the first electronic synthesiser technology to be commercially available and was epitomised by the Mini-Moog and VCR3 synthesisers that were used in the late 1960s and early 1970s. Essentially a set of analogue oscillators and filters that allowed a simple analogue wave form such as a sine, square, or triangular wave to be filtered and modulated to create new waveforms. The most frequently used filter would change the waveform's envelope and this combined with other wave forms such as noise allowed quite revolutionary sounds and effects to be created.

The control medium that is used is an analogue voltage and this led to several problems of instability and oscillator drift. The filters and oscillators were voltage controlled and therefore any change in the control voltage due to temperature or electronic component aging would cause the frequency to drift. With a single synthesiser, this is not necessarily a problem but with the instrument playing with other mechanical and more stable instruments, this could lead to discord as the synthesiser went out of tune. It was not uncommon for several systems to be used for live performances: one to be used for the performance, another acting as a spare and a third being checked and tuned.

The programming of the synthesiser was often performed using a patch bay where the input and output of various modules could be linked together. Simpler models would have a relatively small number of patch connections while the more sophisticated systems allowed additional modules to be added as required to expand the system. Such systems could have several hundred connections — all of which have to be recorded to define and recreate the waveform.

Polyphonic synthesizers were the next development and allowed several notes to be played simultaneously to create chords. These effectively consisted of several monophonic synthesisers working in parallel. The keyboard was scanned to decide which note was being played and to which synthesiser it was allocated. The early voltage controlled polyphonic synths were also liable to frequency drift.

Modern analog synthesisers are not voltage controlled and use digitally controlled filters and oscillators instead. This has dramatically improved the frequency stability to the point that it is rarely a problem.

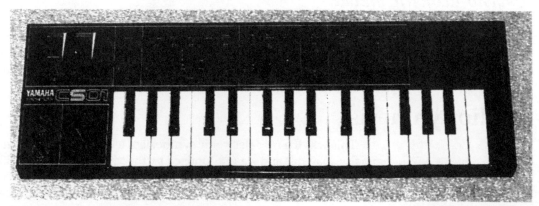

A Yamaha analog synthesiser

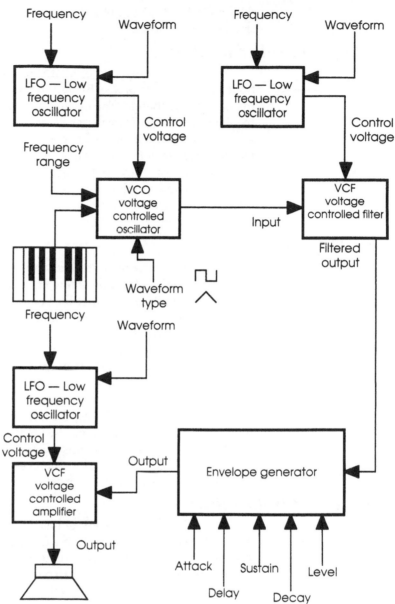

An AM synthesiser block diagram

An AM synthesiser uses a series of voltage controlled audio units to create and manipulate a waveform. Early synthesisers used analogue voltages for control and were notoriously unreliable due to drift and other analogue effects. As technology progressed, the control voltages switched to being digital values, thus providing a more stable platform and the option to integrate the functions onto a single chip. As a result, most AM synthesisers, including the low cost keyboard instruments, effectively have a pre-programmed AM synthesiser at their heart to produce all the different simulations of musical instruments. The block diagram of an AM synthesiser is shown. Each of the basic blocks is voltage controlled and processes or creates a waveform as described.

Voltage controlled oscillator

A voltage controlled oscillator has its output frequency controlled by an input voltage. By varying the input voltage, the output frequency will change. The oscillator normally supports a range of regular output waveforms, such as square wave, sawtooth, sine, and so on. To provide a finer level of control, they can have a range input which determines the frequency range that the input voltage can cover. In addition, they can be connected to a keyboard which can be used to generate predetermined frequencies to match the musical scale. In such cases, the range switch is normally arranged to operate as an octave switch.

Low frequency oscillator

This is a voltage controlled oscillator which has a very low frequency range — typically a few Hz or less. It is normally used as a controlling voltage for other blocks to create vibrato and tremolo effects. Vibrato is achieved by adding the output voltage to the VCO so that its frequency varies by the frequency of the LFO to simulate the vibrato that many musicians use when playing instruments. By varying the frequency and the output level of the LFO, the vibrato effect can be reduced or increased as required. The amplitude of the VCO output remains constant.

The tremolo effect is very similar to vibrato except that the LFO feeds a voltage controlled amplifier where the amplitude is modulated and the frequency remains constant. Incidentally, the tremolo arm on many guitars actually changes the frequency and not the amplitude — it should really be called a vibrato arm!

Voltage controlled filter

A voltage controlled filter has its characteristics controlled by an input voltage. For example, the frequency response can be determined by the input voltage; by connecting the output of a LFO to this input, the frequency response can be cycled. This produces an effect similar to that of a wah-wah pedal.

Envelope generator

This is probably the most important block as it is used to shape the sound. Most natural sounds do not have a consistent amplitude and exhibit many different characteristics. Some sounds have a high amplitude initially and then fade away. Others start quietly and get louder, and so on. The envelope generator defines how the amplitude is shaped and can thus achieve a more natural sound.

Normally, some reference point is used to start the envelope generator and with most synthesisers this is done by pressing a key on a keyboard or by a PC writing a bit into a register. This is used to identify three faces: the starting point (when the key is first pressed), the hold time (the length of time the key is pressed) and the release point (when the key is released). These timing points are used to trigger the envelope generator and with its associated inputs define the envelope waveform.

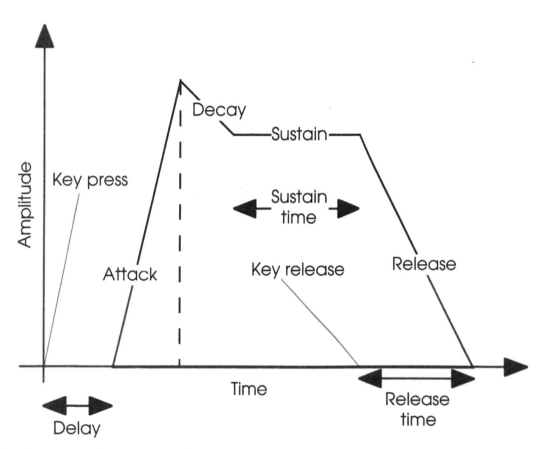

Envelope generator components

Delay   This is the delay — if any — after the key has been pressed and the start of the envelope generation.

Attack   This is the positive slope of the waveform and determines the time for the sound to peak.

Decay   This is the slope of the envelope that determines the time it takes from the amplitude peak to reach the sustain level.

Sustain   This determines the amplitude while the key is pressed down.

Release   This slope determines the length of time it takes for the sound to die out once the key has been released.

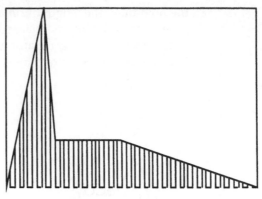

Fast attack, fast decay
low sustain, long release

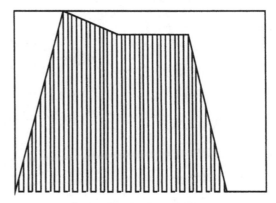

Fast attack, slow decay
high sustain, short release

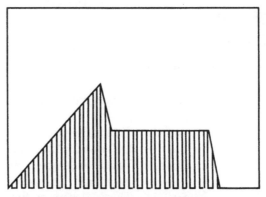

Medium attack, fast decay
high sustain, short release

Example envelopes

Voltage controlled amplifier

This is an amplifier used with an LFO to create tremolo effects. It effectively modifies the envelope but does so using a regular waveform as shown in the diagram.

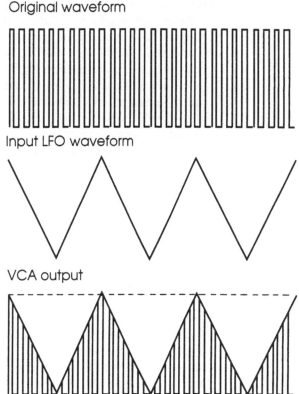

Original waveform

Input LFO waveform

VCA output

VCA operation with a LFO input

FM synthesis

While AM synthesisers can produce a variety of sounds, they tend to lack the realism or depth of sound that instruments and natural sound provide. The reason is that the original sound comprises several different tones which cannot be reproduced by a simple AM synthesiser. As a result, a new type of synthesiser appeared which not only provided a wider range of effects but also simulated instruments far better than anything previously created.

The basis for the new synthesis is the Fourier transform which states that by performing a mathematical operation on a waveform called a fast Fourier transform or FFT, the waveform can be decomposed into several pure sine waves of differing amplitude and frequency. When these constituent sine waves are played back, they combine to create the original waveform.

To create a synthesiser that uses these ideas, several sources or voices are used and combined to create the final output. These sources were initially VFO producing regular waveforms. At this point, digital versions of the analogue synthesisers were appearing that performed the various control operations digitally; it was only a short step for the first digital FM synthesisers to be created. The waveforms were digitised and processed to create the required waveforms

While improving the quality and range of the sounds, it still was not right. However, with a digital waveform source, the actual waveform does not have to be regular and can be a digitised section of analogue sound. This was the big break that was needed. By PCM encoding real instruments and using these as the samples for the voices, the harmonic content could be varied over time and thus a more authentic sound achieved. The initial synthesisers that used this technique utilised 8 bit samples and, whilst an improvement over anything preceding them, still sounded a little synthetic. With better sampling — increased sample size and sampling rates — and processing, this has disappeared to the point where a digital synthesised piano using sampled waveforms sound just like the analogue and original piano.

Having multiple voices also allows multiple instruments to be gathered together to create groups or mini-orchestras. This again was beyond the capability of the AM type of synthesiser. Yamaha were the first to successfully introduce FM synthesisers commercially with the advent of the DX-7 keyboard. This combined both sine wave and sample sounds to provide a selection of sound sources. Multiple selections could be made and layered to create complex sounds. The instrument was polyphonic and different sounds could be allocated to different parts of the keyboard and thus allow the instrument to create an orchestral sound with multiple instruments playing simultaneously. This basic approach has formed the favoured approach for any commercial synthesisers ever since. Several other types of FM synthesis have appeared commercially over the years.

Linear arithmetic synthesis

This method is frequently used in Roland synthesizers and combines both FM synthesis and AM synthesis techniques. The starting point is a high quality sample that contains the first part of the waveform i.e. the attack portion. This is then combined with a synthesised ending using envelope generators and filters. Many samples can be layered together to create different starting points.

Additive synthesis

This is a variation on the standard FM synthesis technique. With FM synthesis, the different voices are digitised, mathematically combined by a processor and the resulting digital data converted to an analogue signal to create the final output. With additive synthesis, sine waves are combined using analogue techniques. This method has failed to gain much commercial acceptance, primarily becuase of the difficulty in programming or editing the sound.

Wave synthesis

This is probably the most commonly used FM synthesis technique, primarily because of the excellent quality of the sounds that can be produced as well as the fact that nearly all PC systems now support or use the technique. The technique has already been hinted at and uses digital samples of sound sources such as musical instruments, standard wave forms and even from old synthesizers. Once sampled, these sound waves can be used as alternatives to sine waves and combined and layered to create both complex and realistic sounds.

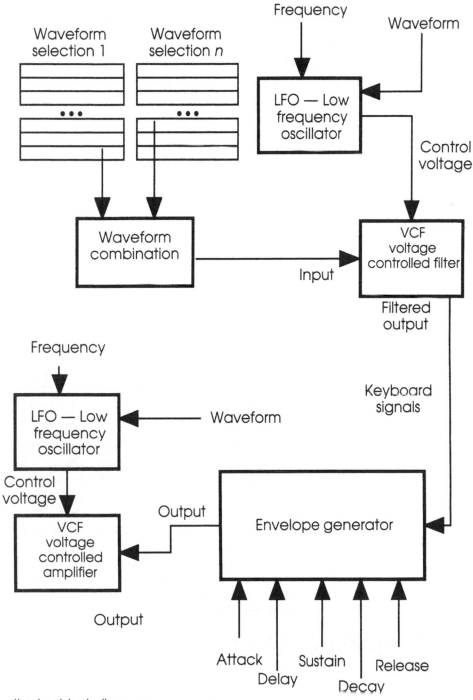

FM synthesiser block diagram

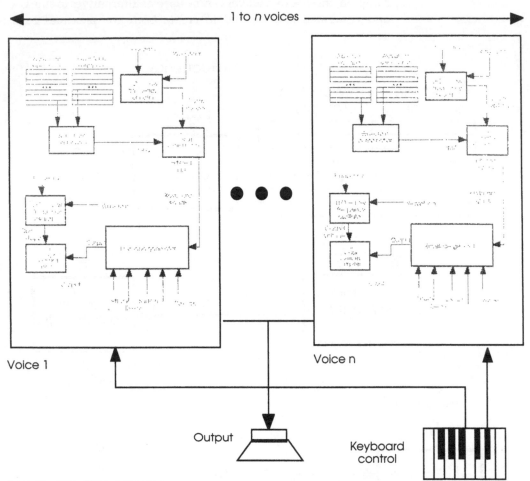

A multi-voice FM synthesiser

A variation on this idea was the development of the sampling synthesiser that allowed users to record their own waveform as the basis of the voice. The processing to change the frequency and other attributes was then used to change the sound or allow the sound to form the basis of a voice. This technique could be used to record speech, crashes or other noises and play them back at any frequency. By lengthening the sampling time, whole sections could be digitally recorded for playback or as a basis of a voice. This sampling technique has become widespread to the point where copyright infringements exist where one musician's work has been directly copied and used by someone else without permission.

Digital signal processing

With sound samples digitally recorded, it is possible to use digital signal processing techniques to create far better and more flexible effects units (or sound processors, as they are more commonly called). Such units comprise a fast digital signal processor with high resolution (16 bit) A to D and D to A converters and large amounts of memory. An analogue signal is sent into the processor, converted into the digital domain, processed using software running on the processor to create filters, delay, reverb and other effects before being converted back into an analogue signal and being sent out.

These units are derivatives of synthesisers and have a lot in common with them. They can be completely software based, which provides a lot of flexibility, or they can be pre-programmed. They can take in analogue or, in some cases digital data, and feed it back into other units or directly into an amplifier or audio mixing desk, just like any other instrument.

Creating echo and reverb

Analogue echo and reverb units usually rely on an electro-mechanical method of delaying an audio signal to create reverberation or echo. The WEM Copycat used a tape loop and a set of tape heads to record the signal onto tape and then read it from the three or more tape heads to provide three delayed copies of the signal.

A modern digital delay unit. This is not much bigger than a pack of cards and is about 1/50th of the size of a WEM Copy cat.

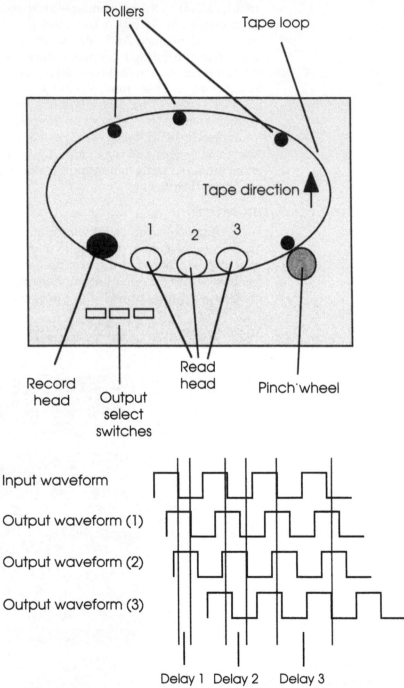

A tape loop based delay unit

The delay was a function of the tape speed and the distance between the recording and read tape heads. This provides a delay of up to 1 second. Spring line delays used a transducer to send the audio signal mechanically down a taut spring where the delayed signal would be picked up again by another transducer.

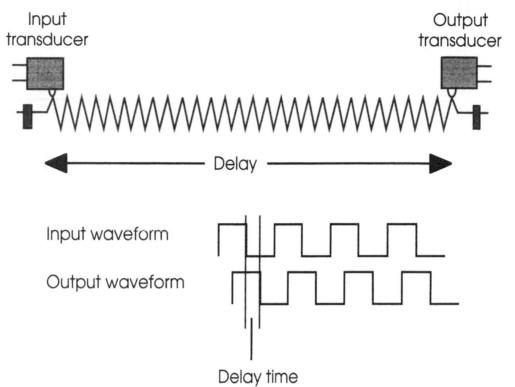

Input
transducer

Output
transducer

◄————————— Delay —————————►

Input waveform

Output waveform

Delay time

Spring line delay

Bucket brigade devices have also been used to create a purely electronic delay. These devices take an analogue signal and pass it from one cell to another using a clock. The technique is similar to passing a bucket of water by hand down a line of men. Like the line of men, where some water is inevitably lost, the analogue signal degrades — but it is good enough to achieve some good effects.

With a digitised analogue signal, creating delayed copies is easy. The samples can be stored in memory in a buffer and later retrieved. The advantage this offers is that the delayed sample is an exact copy of the original sound and, unlike the techniques previously described, has not degraded in quality or had tape noise introduced. The number of delayed copies is dependent on the number of buffers and hence the amount of memory that is available. This ability, coupled with a signal processor, allows far more accurate and natural echoes and reverb to be created.

The problem with many analogue echo and reverb units is that they simplify the actual reverb and echo. In natural conditions, such as a large concert hall, there are many delay sources as the sound bounces around and this cannot be reproduced with only two or three voices which are independently mixed together with a bit of feedback. The advantage of the digital approach is that as many delays can be created as required and the signal processor can combine and fade the different sources as needed to reproduce the environment required.

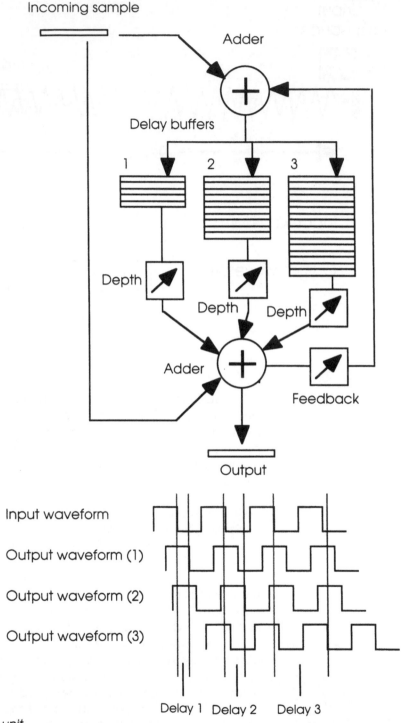

A digital echo/reverb unit

The block diagram shows how such a digital unit can be constructed. The design uses three buffers to create three separate delays. These buffers are initially set to zero and are FIFOs — first in, first out —thus the first sample to be placed at the top of each buffer appears at the bottom at different times and is delayed by the number of samples that each buffer holds. The smaller the buffer, the smaller the delay. The outputs of the three buffers are all individually reduced in size according to the depth required or the prominence that the delayed sound has. A large value gives a very echoey effect, similar to that of a large room or hall. A small depth reduces it. The delayed samples are combined by the adder with the original sample — hence the necessity to clear the buffer initially to ensure that random values, which adds noise, do not get added before the first sample appears — to create the final effect. A feedback loop takes some of the output signal, as determined by the feedback control, and combines it with the original sample as it is stored in the buffers. This effectively controls the decay of the delayed sounds and creates a more natural effect.

This type of circuit can become more sophisticated by adapting the depth with time and having separate independent feedback loops and so on. This circuit can also be the basis of other effects such as chorus, phasing and flanging where the delayed signal is constantly delayed but varies. This can be done by altering the timing of the sample storage into the buffers.

IBM PC sound cards

It should not come as a surprise that the PC has taken advantage of all the developments in digital audio and has combined them in a single card to give several ways of reproducing and creating audio.

While the Apple MAC has provided sound support right from its inception, it has taken several years for the IBM PC world to catch up and provide similar functionality. Although the IBM PC has always had a speaker that could be used for play back, it never had the hardware support in terms of analogue to digital and digital to analogue converters to allow its use to play back and record sounds. While the early MACs were limited to playback, this essential support was always there and has been expanded to include stereo support and recording. To respond to these developments, many third parties started to develop sound cards for the IBM PC and in particular to support Windows and PC games. The IBM PC method of providing sound support offers the ability to play back and record sounds, provides various sound synthesisers and MIDI support for the control of external digital musical instruments. Many such cards also provide interfaces for CD-ROM support.

The most well known IBM PC sound card today is the SoundBlaster from Creative Labs. It can create and play back PCM encoded sounds and store them as WAV files for use later as replacements for operating system beeps, and so on. It has a synthesiser and a MIDI interface to allow it to be controlled by external keyboards and allow the PC to control them. In this way, the hardware provided by the card, coupled with software to run on the PC, allows the PC to become a synthesiser, effects unit, or the basis of a sophisticated music system controlling and programming MIDI based instruments.

Anatomy of a sound card

Sample size and sampling rate

Sample size is a good indication of the cost and quality of the sound card. Early versions used 8 bit sizes while today 16 bit sound cards are the norm, with the top range boards supporting 32 bit data. The more data, the higher the quality of sound — but the amount of data dramatically increases. To keep the data storage down to a minimum, data compression is often used.

A second associated parameter is the sampling rate. This determines how many samples are taken and again, the higher the value, the better the quality but the more data storage is needed. The top range boards can provide CD quality sound using 16 bit samples at a sampling rate of 44.1 kHz.

Synthesiser

The synthesiser portion of a sound card can create sounds or voices using synthesiser techniques in much the same way as a synthesiser keyboard can. They can often support multiple voices and it is possible to create a complete orchestra using the card. Two techniques are commonly used to synthesise sounds: FM synthesis and wave table synthesis. FM synthesis uses frequency modulation of carrier waves to create the sound and, with the addition of an envelope generator to control the attack and sustain of the sound, can create a wide range. The second technique uses tables of digital information sampled from real instruments and sound sources as its basis for synthesis. It provides a more natural sound but requires large amounts of storage and processing power to offer the range of sounds when compared to FM synthesis. It is, however, gaining popularity because of the quality of the sounds it can produce.

A third method is also appearing which is taking the synthesiser module from commercial keyboard synthesisers and adding them to the sound card. They can either be controlled directly or through MIDI.

The synthesiser section is also used to provide a MIDI instrument and allow the playback of MIDI music without connecting to an external instrument.

Some boards, such as the SoundBlaster 16, have an interface that allows other synthesisers to be added to the sound card using a daughter board. Yamaha provide a card called the Wavetable that provides this.

Audio mixer

With all the different sound sources and only a single pair of speaker outputs, an analogue mixer is often supplied to mix the different sounds so that they all can be heard.

Digital signal processor

This provides the conversion of analogue sounds into digital format, and vice versa, and allows the sound to be processed and stored. This includes data compression of digitised audio. It is this part of the card that allows the sound card to record and playback sounds from the PC. More expensive cards provide very sophisticated facilities including surround sound and effects like reverb and echo.

MIDI

The MIDI port allows the control of external digital musical instruments, such as synthesisers, sequences and effects units from the PC. Coupled with editing software, it is possible to use the PC as a musical word processor and compose music. A special lead is often needed from the port to provide the optical isolation that is needed. The port cannot and should not be connected directly to a MIDI device. This topic will be covered in more detail at the end of this chapter.

Joystick

Most sound cards provide a joystick port which doubles as the MIDI port. If the port is used as a joystick port, it must be enabled and any other joystick port within the system must be disabled to stop any conflicts or similar problems. The enabling is normally done through a jumper setting.

CD-ROM

Most sound cards today have at least one CD-ROM interface. For more information, refer to Chapter 6 on CD-ROMs.

Installing a sound card

Installing a sound card is relatively straightforward, providing the installation procedures supplied with the card are followed. Having said that, I have received cards where the instructions were a little confused (to say the least!) and the successful installation was thus not easy.

Hardware settings

Like any expansion card, setting the board correctly so that it does not use the same interrupt, DMA channel or port address of another card in the system is key. Most boards have a set of jumpers to configure the board and then require those settings to be input into a configuration program to set up and configure the drivers. The Creative Labs SoundBlaster card — probably the most well known of the myriad of cards available — has an installation program that can auto detect the hardware settings if they are unknown.

Test the card

If the card comes with any test software, it is worth using it to check that the speakers are connected correctly and not inserted in the microphone socket by mistake, for example, and that the volume control is turned up.

MS-DOS drivers

The next step, assuming the card tests correctly, is to install the software drivers. The first set is normally installed for MS-DOS and will often require hardware settings.

Windows drivers

Installing Windows drivers can be a little tricky, depending on the card, and whether special Windows installation software is supplied. With the SoundBlaster card, this is the case and installation is as simple as using Program Manager to run the installation program. For other cards that are compatible with one of the Windows supported cards, SoundBlaster, ADLIB and so on, the driver is installed from within Windows by opening the drivers control panel and adding the required driver. As part of the set-up, the hardware settings of the card will be requested.

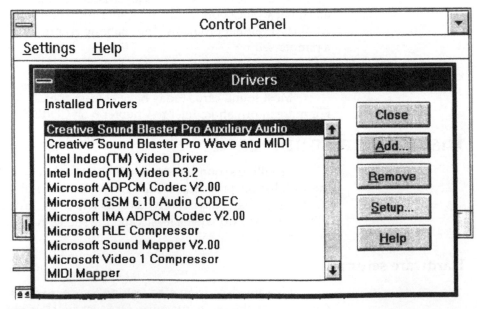

Adding a Windows sound card driver

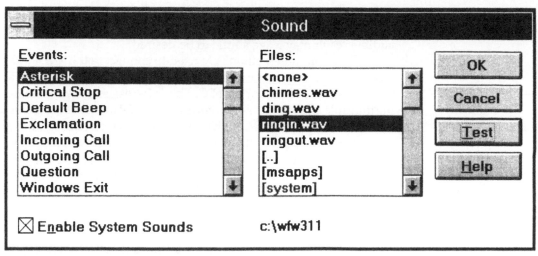

The Windows Sound control panel

Once the driver has been installed, it can be tested using the Windows Sound control panel. This allows sounds to be selected for certain Windows operations and by clicking on the test button, the selected sound can be played back.

MIDI

MIDI (musical instrument digital interface) is a standard way of passing digital musical information from one MIDI device to another. It has become the standard for anyone interested in combining computer technology and music.

The interface is essentially a serial one which is clocked at either 512 kHz or 1 MHz and sends data around a daisy chained set of devices. The data is sent digitally with the description of musical notes, keyboard presses and timing built in as a series of small compact messages. Devices can be set to respond to any message or, by assigning them to a channel, messages can be sent to them alone. Control information can also be sent and, by using a system exclusive message, a manufacturer can define his own set of messages and codes. This allows the settings from a synthesiser of effects unit to be stored on disk on a PC for future reference, and so on.

Within the context of an IBM PC or Apple MAC, MIDI is used to allow the computer to become the controlling device within a MIDI configuration. In this way, the computer combined with its ability to store and process data, can take a musical composition played on a MIDI keyboard and convert it into musical notation on the screen. Alternatively, it can be used to generate the music and then send the sequences to a MIDI device to play them back. This has led to the widespread exchange of MIDI files which playback music, either original or covers of popular songs and tunes. This is a very compact way of sending audio but it does rely on consistent MIDI channel allocation. If the playback machine allocates channel 5 to a percussion instead of a piano, the MIDI file will play correctly except that the audible result may leave a little to be desired!

Two common methods are used to connect to the MIDI interface. The first uses the multiple MIDI out ports to connect a second synthesiser to the MAC or IBM PC. This is the easiest way of connecting multiple MIDI devices—providing the MIDI interface has enough OUT ports to support this. With most MAC interfaces, this is the case.

With the IBM PC, this is not always so. In this case, the second example shows how the problem can be overcome. It uses the MIDI through port on the keyboard to provide the link. In both cases, the computer will accept data from the keyboard and transmit data to either synthesiser.

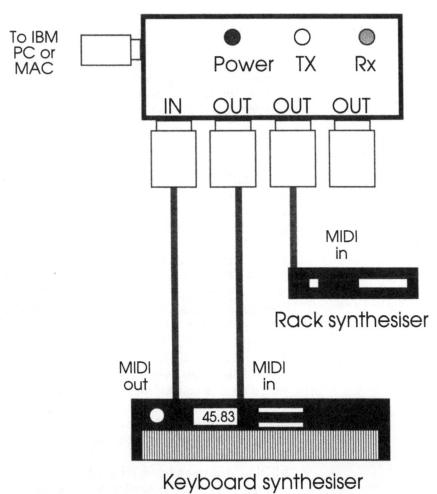

Example MIDI configuration — 1

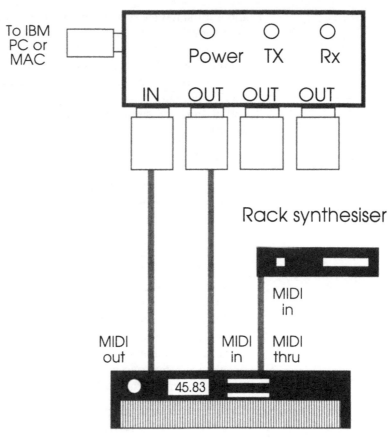

Example MIDI configuration — 2

Using IBM PCs with MIDI

The first obstacle to overcome is the correct configuration of the MIDI port itself: it must not be configured or used as a joystick port

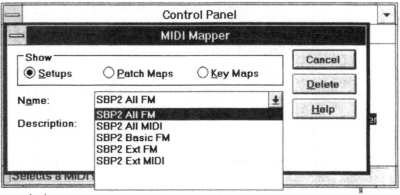

Windows MIDI mapper control panel

and it must have a special lead that ends in the 5 pin DIN connector used for MIDI systems.

The lead is usually be purchased separately as it is not often supplied as standard. The second point is that Windows will often map the MIDI data to go to the sound card on board synthesiser. To check where the data is going, open the MIDI mapper control panel and change the settings to 'ext MIDI' to access an external MIDI device.

Using Apple Macs with MIDI

One fast growing application area is that of MIDI systems, where a MAC can be used to control a range of synthesisers, sequencers, drum machines and other musical instruments through the use of a MIDI interface.

The first step is to buy an interface box. This plugs into either the printer or the modem port, providing a single MIDI in port and, usually, three separate out ports. Many MIDI interface boxes are available but some compatibility problems exist. The first concerns the interface's suitability for use with MAC Portable and PowerBook systems. These models can turn off power to the SCC devices which can cause problems with some MIDI interface boxes. The second problem is concerned with the fact that a printer or external modem can no longer use the port that the MIDI interface is plugged into.

Typical MIDI set-up panel

Some interfaces provide a switch and a second mini-DIN socket so that the original peripheral can be plugged in and the MAC switched between the MIDI interface and the peripheral via a switch. Most modern designs take power directly from the MAC and optically isolate the MIDI devices from the interface. Be wary of interfaces that need external power supplies.

The SCC ports in the MAC IIfx are intelligent and may require special software drivers to allow the interface to work correctly. All in all, the best advice is to make sure that the interface has a switched through socket and definitely works with your particular MAC model.

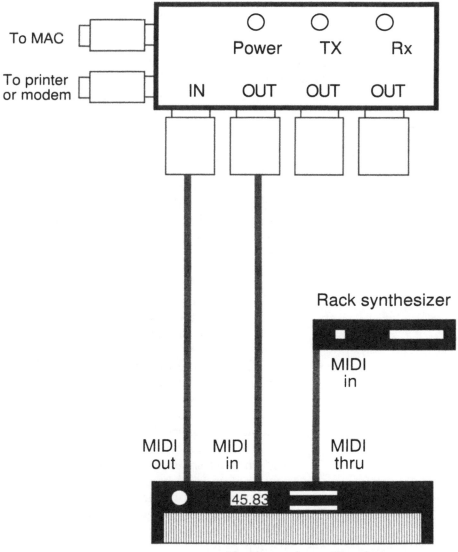

A MAC MIDI configuration

There are some software issues which may catch out the unaware. With MACs fitted with internal modems, such as the Portable and PowerBook, make sure that the internal modem is disabled. This is done for the portable Macs through the use of the Portable control panel. If the interface is connected to the printer port, make sure that any AppleTalk link to a LaserWriter or network is disabled. I once spent several hours attempting to get a MIDI interface to work before discovering that this was the problem. AppleTalk data was driving the status lights on the interface so that the unit appeared to be working and transmitting MIDI data. Unfortunately, none of the MIDI instruments connected to the interface would work as they did not understand the AppleTalk data or protocol.

Many MIDI applications normally assume that the modem port (and not the printer port) is being used, although most will offer some way of swapping to the other port. Others also offer a 'MIDI through' type function by copying data that arrives at the MIDI 'in' port to the MIDI 'out' port. In this way, a keyboard can control several other synthesisers while still retaining the ability to input data to the MAC.

Summary

The PC of today can handle digital audio extremely well with the addition of a sound card. It can convert analogue sound to and from the digital domain and, once digitised, use it to great advantage. The provision of synthesisers and MIDI interfaces is a great step forward because it allows good quality music audio to be generated without having to digitise and store megabytes of PCM encoded sound.

- PCM encoding with compression
  This technique is used to record speech and other non-musical sounds and store them as .WAV files. Once in this format, they can be processed to create almost any type of effect needed, providing sufficient memory and processing power are available. Whilst it can be used to store music, it is less efficient if the main interest is in the tune as opposed to the definitive sound. An Eric Clapton guitar solo can be converted to a MIDI file, but although the notes and tunes would be reproduced, the actual sounds that makes an Eric Clapton solo what it is, would be lost.
  PCM encoded sound is the fundamental starting point for many multimedia effects and is frequently used for small sound snippets. Providing there is sufficient storage available, it can be used to completely replace the analogue recording tape. Such systems are becoming available but they are expensive at present.
- FM synthesiser
  Programming a synthesiser is another good way of producing audio but it tends to suffer from a lack of standardisation.

- FM Synthesiser with MIDI
  This is fast becoming the way of transferring and playing musical soundtracks on PCs for games and other multimedia software. It is compact but does rely on having a MIDI compatible synthesiser installed in the PC and correctly configured. Fortunately, with the advent of the sound card, this is usually the case and this restriction is not a problem.

3 Digital video

This chapter describes the techniques and principles behind the display of video pictures in multimedia systems. It covers how the video information is digital and then translated into analogue signals for display on a monitor or TV set, via its SCART interface.

There is a single key operation that any multimedia system has to perform which makes it different from a normal PC: it has to be able to insert the video into the existing screen display, often into a special window, and still let that window be moved and changed in size as if it was a 'standard' PC window. As will be described, this can be a little tricky, especially with the more sophisticated video options that PCs now support, such as the ability to change picture resolution and colour depth on the fly.

Building a digital picture

The picture displayed on a TV or PC screen is composed of picture elements called pixels, with each pixel containing information about the colour that the pixel should show. Pictures on the screen are made up from a mosaic of pixels and the picture appears as a continuous unit because of their small size.

Pixels and colour depth

The sets of horizontal pixels are known as lines. This term came from the television industry, where the picture height is often referred to as so many lines.

In the PC industry, the dimensions of a picture are normally defined as x pixels by y pixels, where x is the number of horizontal pixels and y is the number of vertical pixels (or lines) in the picture.

Each pixel is stored in a memory array with a storage element representing it. The amount of data stored per pixel determines the number of grey scales or colours that can be displayed. This is known as the colour depth. With a monochrome picture where the pixels are stored as a single bit, only two colours (i.e. black and white) can be displayed. With a 24 bit colour picture, 2 to the power of 24, i.e. over sixteen million colours, can be shown.

The colour depth can have a dramatic effect on the realism of the picture being displayed. The smaller the colour depth, the less realistic the picture will be and, correspondingly, the poorer the resolution will appear to be. The greater the colour depth, the more realistic the picture is and the resolution will be perceived to be far higher.

The pictures shown to demonstrate this are taken from a PhotoCD, and have had their colour depth electronically reduced. As the depth is reduced, artefacts, such as banding, appear. Banding is where a continuous colour change is reduced to only a few colour changes due to the reduced colour selection. As a result, the once continuous change is replaced by a few bands of colour. In the picture the sky and courtyard exhibit this effect.

Pixels

The allocation of pixel bits is normally designed to fit with an RGB colour representation. This is where the pixel bits are divided into three sections, each section being allocated to a particular primary colour: red, green or blue (hence RGB). With a 24 bit pixel, this is straightforward. Each colour component is allocated 8 bits. With an 8 bit colour depth, the allocation is unequal and the concept of palettes is used. This will be explained later.

Bit planes

Within a PC, the memory used to store the pixels is referred to as video memory or VRAM. In terms of structure, it is organised as a set of bit planes, where similar data is collated together. The diagram shows the basic principles. The pixels are organised into a two dimensional map or plane. A third dimension can be visualised behind each pixel that can represent another attribute. If this is colour depth, it can be split into three components representing the bit allocation to each primary colour. In this way, these bits form other planes as shown. There are red, green and blue bit planes within the pixel storage.

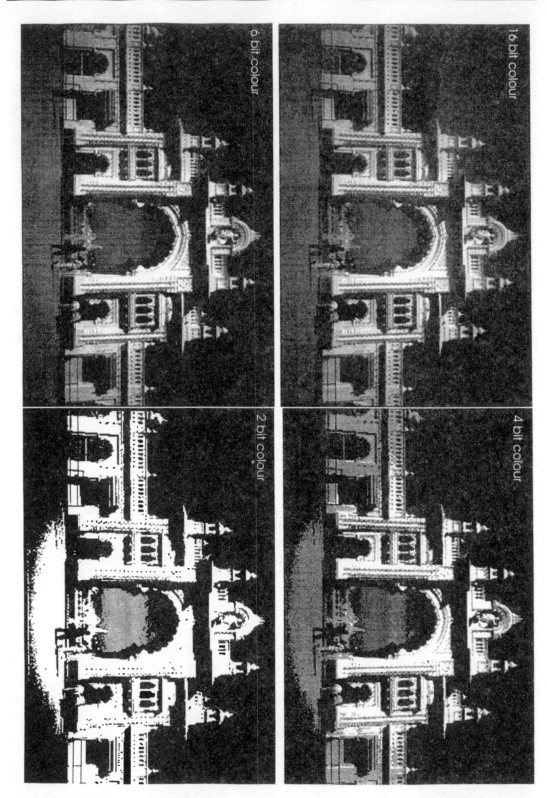

16, 6, 4 and 2 bit colour depths

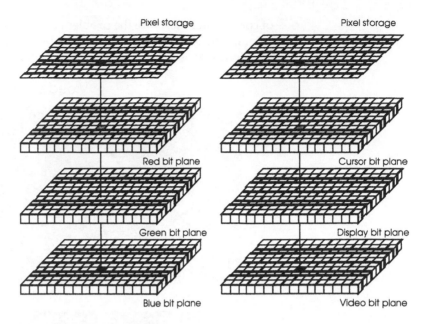

Pixel storage   Pixel storage

Red bit plane   Cursor bit plane

Green bit plane   Display bit plane

Blue bit plane   Video bit plane

Bit planes

This idea of planes does not have to be restricted to colour depth, as the second example shows. Here, the bit planes are defined as picture types, where the end display is composed of data from the display, cursor and video bit planes, which are then combined to create a picture.

Such bit planes can provide data which is combined or masked with other control bit planes to allow parts of the lower bit plane to be visible. The example shows how part of a video image can be masked so that only a portion appears. The video bit plane is combined with the masking bit plane so that only video pixels which correspond to the white masking pixels are displayed. This concept of using bitplanes for control is very important.

There is one important thing to note: the memory organisation of the pixel data and colour depth is variable and depends on several trade-offs. PC memory is organised around bytes and it is useful to also have the colour depth arranged in this way. For control bit planes, bits are often all that is required to provide an on-off control. As a result, these do not fit well together. One option is to assign both the pixels and mask plane one byte per entry — but this is inefficient and wasteful for the masking plane. Another option is to pack the masking bits so that each byte contains the control information for eight pixels. The organisation of the pixels and bit planes within the memory is important and does vary depending on how fast the graphics need to be performed and how little memory is needed.

Palettes

It was mentioned earlier that with low and medium colour depth graphics, the distribution between the various primary colour planes is unequal because the data size allocated to each pixel is not a multiple of three. Most monitors are capable of displaying millions of colours and therefore the problem arises over deciding which 256 colours out of these millions should be allocated to the pixel. This decision also has to consider how the colours are going to be allocated.

Video bit plane

Masking bit plane

Display

Masking and video bit planes

This is where the idea of a palette comes in. A palette chip is essentially a look-up table with a set of entries — usually 24 bits in size — which contain the actual digital values sent to the digital to analogue video converters (DACs) that transform the pixel colour depth information into an analogue value for the red, green and blue inputs on the monitor. The DACs are normally capable of at least 8 bit resolution and, with one per primary colour, this supports 24 bit colour. The palette chip takes the input pixel, uses the value to index into its lookup table and then outputs the table entry. The entries reflect the colour choice needed — i.e. the palette. In this way, the choice of colours is not restricted by the colour depth and can be changed. For work on a sunset picture, more reds and yellows can be allocated. For grey scale pictures, the palette can be set up to provide all shades of grey.

The advantage of this is that the resulting pictures seem to be more realistic, unless a wild assortment of colours is used. A choice of 50 blues to represent the sky instead of 16 allows a far more photo realistic picture to be displayed.

Many graphics file formats now allow a palette to be encoded with the pixel data to allow the palette to be set up accordingly when the picture is displayed. Many PC based drawing and painting packages allow the colour shades to be modified. What is actually happening is that the software is programming and manipulating the palette entries to create the new hues.

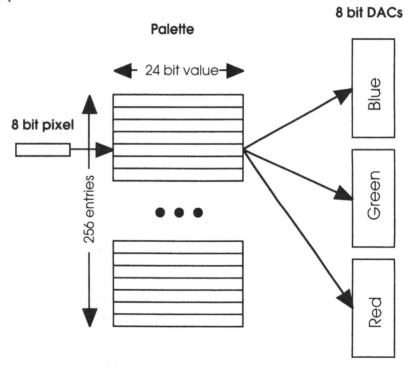

Colour palette

A PC graphics controller

Given the basic principles so far discussed, how does a PC display graphics on a screen? The block diagram for a generic PC card is shown overleaf. It consists of several controllers, a memory array to hold the pixel and colour depth data and three DACs.

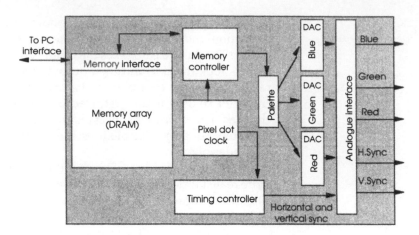

*Generic PC graphics adapter
(DRAM)*

Memory

This is where the pixel data is held and manipulated. In this block diagram, it is assumed that the PC itself is controlling the graphics and thus changing the pixel data as objects are moved on the screen. In low-cost adapters, this memory is usually DRAM i.e. the same as used for the main PC memory. Some PC designs even use part of the main memory for their video store.

Pixel dot clock

This is the key timing clock. It is used to access pixels from the store and provides the fundamental timing reference used to generate the horizontal and vertical synchronisation pulses, and thus feed both the memory controller and timing controller.

Memory controller

This is responsible for accessing the pixel information on each dot clock transition and supplying it the palette chip for palette conversion. It is usually a programmable device that allows different pixel organisations to be supported within the memory array.

Palette

This provides the palette function and receives the pixel data from the memory controller. This data is then converted through the use of a lookup table to create the paletised pixels. These are then sent to the DACs.

DACs

These convert the paletised pixel information into three analogue signals, representing the red, green and blue colour information.

Timing controller

This device takes the dot clock and generates the horizontal and vertical sync signals that control the scanning on the monitor.

Analogue interface

This takes the colour and sync information and provides the analogue signals. Typically this involves the use of transistor drivers to supply sufficient power to provide a good signal into the 75 Ω impedance normally used by the monitor interface. The diagram shows the signals as separate but sometimes they are combined. For example, composite sync, where both the horizontal and vertical sync signals are combined, and sync on green, where the sync signals are supplied on the green colour channel, are two common examples.

Using video RAM

One of the problems that the previous block diagram has is memory contention between the PC and the graphics memory controller. Both need to access the memory. The PC needs access to manipulate the images in response to the user: pull down menus, highlight text, move the cursor and so on. The memory controller needs access to supply the pixel information. As a result, the contention over the shared interface results in delays with the graphics side usually having priority.

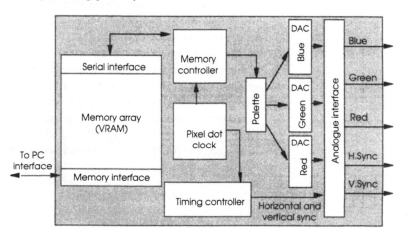

Generic PC graphics adapter (VRAM)

A solution to this is to use video RAM or VRAM. This is a special form of DRAM that has a serial interface and thus allows data to be accessed simultaneously by the PC and the graphics side.

The serial port works by transferring a row of data into a buffer which can then be clocked out as needed. While this is going on, the PC can access the memory using the normal address and data bus interface. As VRAMs are more expensive than DRAM, low cost cards tend to use DRAM while higher performance cards use VRAM.

RGB and YUV colour representations

Up to now, the picture data and its colour information have been described in terms of its three primary colour components. This is not the only way of representing this information, although it is the most commonly used technique with PCs.

An alternative and one that is used with television pictures is to convert the picture into luminance and chrominance components. Luminance contains the grayscale information and can be used to display a black and white version of the picture. The colour information is encoded in the chrominance signal. The conversion and descriptions of the different formats are in the next chapter.

Frames and interlacing

Frames and frame rate

With a moving picture or video, a third dimension is also applicable to the simple two dimensional x and y size of the picture — the frame rate. Each picture is called a frame and multiple frames must be displayed in sequence to create the illusion of a moving picture. The frame rate defines how many frames are displayed per second and this value usually ranges from about 24 to 60, although higher frame rates are appearing or being experimented with. The minimum frame rate needed for motion is about 24 frames per second (fps) and this is the standard used for movies.

Given this, the bandwidth needed to display and transmit a picture must be considered. If the number of pixels per frame is taken and multiplied by the frame rate, this gives an approximate value for the bandwidth needed. It is approximate because the bandwidth for displaying a picture must include the areas that are not displayed i.e. those used for flyback (see later) and so on. This bandwidth value is also a good indication of the speed needed for the dot clock (see later).

640 x 480	@ 25 fps	=	7.68	MHz
1280 x 960	@ 25 fps	=	30.72	MHz
640 x 480	@ 50 fps	=	15.36	MHz
1280 x 960	@ 50 fps	=	61.44	MHz
640 x 480	@ 75 fps	=	23.04	MHz
1280 x 960	@ 75 fps	=	92.16	MHz

Approximate bandwidth requirements for different frame sizes and rates

Interlacing

Interlacing is a technique used with televisions and some PC monitors to overcome bandwidth limitations. If there is insufficient bandwidth to support the required frame rate, the persistence of vision of the eye can be exploited to reduce the amount of data sent in each frame. Instead of sending a complete frame, half the frame is sent — reducing the bandwidth requirement by half.

Original image

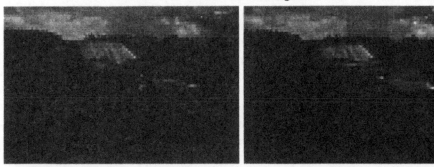

Interlaced images

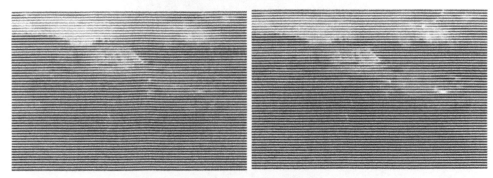

De-interlacing: duplication   De-interlacing: duplication

De-interlacing: interpolation   De-interlacing: interpolation

De-interlacing techniques

The frame is split into two by removing alternate lines to create two fields, one consisting of odd lines and the other even lines. The two fields can be joined together to create the original picture. By sending alternate fields, the overall frame rate is dropped to half but is still sufficient to fool the eye that the picture is continuous. Persistence of vision now not only links the frames together into a seamless picture but also recombines the odd and even fields so that the viewer is normally not aware of the interlacing.

Full

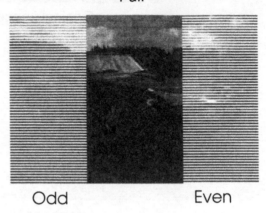

Odd   Even

Interlacing

Interlacing also provides an opportunity for video compression by losing one of the fields and re-creating the missing field from the data in the first one. This loses data but can achieve reasonable quality. There are several algorithms that do this, the simplest being duplication and interpolation. With duplication, the missing field is recreated by duplicating the first one. In effect, the lines in the first field are simply doubled in vertical size. The interpolation technique mathematically analyses the pixels around each missing pixel to create the missing value. This can be a simple two pixel analysis or more complex, using large matrices. The diagram shows the results of duplication and interpolation through simple averaging. Interpolation gives the best results but requires more processing. Duplication gives a better result than might be expected.

The monitor

Until now, the discussion has focused on the digital side of video and graphics. Most monitors and televisions have an analogue interface and therefore some form of conversion is necessary. With a PC, digital video is converted into a set of analogue signals which are fed into the monitor. With a television set, which is really nothing more than a monitor connected to a sophisticated radio receiver, the video information is transmitted in analogue form and is simply extracted. To do this, it is necessary to understand how a picture is displayed on a screen. This is covered now.

The picture is made by scanning an electron beam horizontally across the screen and modulating its intensity by the pixel values that correspond to its position. The length of the line is divided by the number of pixels that it must display. Each section is then filled in by the electron beam, depending on its intensity as set by the value of the pixel. The timing is quite tight but it can easily be generated from the dot clock. With a colour picture, this arrangement is triplicated with an electron beam effectively for each primary colour.

CRT screen

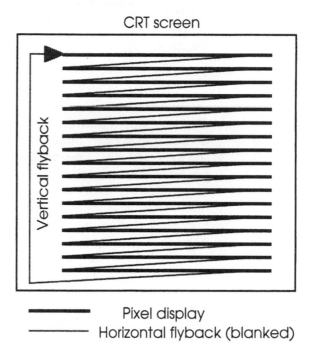

Pixel display
Horizontal flyback (blanked)

Flyback

When the end of each line is reached, the electron beam must be repositioned at the beginning of the next line. This is done by blanking the beam and moving it back to the next line's starting position. This movement is done by the monitor in response to a synchronising signal called the horizontal sync. The beam is blanked during this process to prevent its movement from destroying the previous image. This process of moving the beam to the new position is known as horizontal flyback or simply flyback.

This process is repeated for all the lines until the last one is reached, when the beam is positioned back at the top of the picture by vertical flyback, which is triggered by a vertical sync signal. Both these sync signals can be positive or negative and they can be combined with other signals to create multiple signals on a single line.

Digital to analogue conversion

To create these signals from a digital format is not too difficult if the pixel memory store is artificially enlarged so that dummy pixels are added to the picture and used to create the sync pulses. This is the basis of the technique that is used with many controllers.

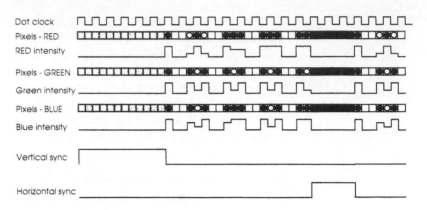

Generating analogue signals

The dot clock is used to clock out pixels and these values are converted using the DACs to the analogue intensities as shown in the diagram. When the last pixel in a line is clocked out, a counter is used to clock dummy data through which is used to blank the intensities and thus create the blanking signal. The dummy data is also used to generate the horizontal sync. Once sufficient time has passed to reach the position defined by the number of dot clocks that the counter lets through, the next line of pixels is sent through. At the end of the last line, a different counter is used to create the vertical sync and allow time for the starting position to be reached.

The system can support interlacing, where alternate fields are used to generate the position and intensity information. This is often used with commercial television systems. The actual analogue values used are obtained by using transistors to provide the correct voltage and impedance values. The advantage of this type of system is that it is totally programmable and thus can support many different resolutions and timings. This has led to a plethora of different resolutions, colour depths and frame timings available on the PC — particularly with the IBM PC.

VGA and SVGA standards

The IBM PC uses a graphics adapter called VGA which supports many different modes. The pinout shown in the table identifies the signals it uses. The three video signals are analogue and range from 0 to 1 volt to provide a variable colour intensity that can support high quality i.e. large colour depth displays. Not all displays can do this and this often separates the good from the bad. The sync signals are kept separate. The standard supports monochrome, which should more clearly be called grey scale, by dropping the red and blue video signals so that only the green signal is used. This does lead to colour duplication where the adapter displays several different colours with the same value green component. With a colour monitor, they appear as different colours. With the monochrome version they appear as the same shade of grey on the screen and this can cause confusion on a display by hiding objects that would normally be visible to the user.

Pin	Monochrome VGA, SVGA & MGA	Colour VGA, SVGA & MGA
1	Not used	Red video
2	Video*	Green video
3	Not used	Blue video
4	Not used	Not used
5	Ground	Ground
6	Not used	Red return (ground)
7	Not used	Green return (ground)
8	Video return (ground)	Blue return (ground)
9	Key pin	Key pin
10	Sync return (ground)	Sync return (ground)
11	Monitor ID (not used)	Monitor ID (not used)
12	Monitor ID**	Monitor ID**
13	Horizontal sync	Horizontal sync
14	Vertical sync	Vertical sync
15	Not used	Not used

Notes:
*   Monochrome monitors use the green video channel as their source of video.
**   If this pin is grounded, the monitor is assumed to be monochrome. If open, it is assumed to be colour.

VGA pinouts

MDA – Monochrome Display Adapter

This was the first display system produced for the PC. It supported a display of 25 rows with 80 characters per row. Each character could either be white, intense white, underlined or reversed. This display was fine for text work but did not allow any access to the individual pixels.

Hercules Adapter

The Hercules card (provided by the company of the same name) quickly became a de facto standard. It is compatible with the MDA standard but adds a graphics mode which allows individual pixels to be accessed. The maximum screen resolution is 748 pixels horizontally by 350 vertically. For monochrome work, it is still a good choice, giving a fairly high resolution graphics screen.

CGA — Colour Graphics Adapter

This was the first IBM display to support colour graphics. It is compatible with MDA text mode, except that each character can be displayed in one of 16 colours and can be made to flash. In graphics mode, it supports a 16 colour palette with a screen resolution of 640 horizontally and 200 vertically. This gives the colour graphics an incredibly 'chunky' look, reminiscent of that produced by early 8 bit games computers on domestic televisions. In addition, the text characters are not as crisp as those produced by the MDA.

Many CGA boards were used with monochrome monitors, where the 16 colours were represented by 16 grey shades – but the appearance of pixels meant that CGA did not really replace the Hercules board unless colour was absolutely essential.

CGA monitors are becoming increasingly difficult to find and their value is not high. Some multisync monitors can support both CGA and VGA/SVGA inputs.

EGA – Enhanced Graphics Adapter

The next attempt was EGA. This is compatible with CGA and MDA but allows a selection of 16 colours to be taken from a palette of 64. The graphics resolution approaches that offered by the Hercules board with a 640 pixel horizontal resolution and a 350 pixel vertical resolution. The text mode was also improved by increasing the number of rows from 25 to 43. Fortunately, the old mode of 25 rows per screen is still supported as 43 rows per screen is cramped, to say the least.

MGA – Multi-colour Graphics Array

This standard appeared on IBM PS/2 models 25 and 30 but it has virtually been replaced by VGA – it is a subset of the full VGA standard. It supports all the CGA functions but adds a 300 by 200 mode, which supports 256 colours, and a high resolution mode that supports 640 by 480 pixels with two colours. One important difference between MGA and its predecessors is the move away from the old 9 pin D type connector and TTL levels to a 15 pin high density D type and analogue signals.

VGA – Virtual Graphics Array

This became the de facto standard for nearly all PCs. It supports MDA, CGA, EGA and MCGA and allows up to 256 colours from a palette of 65,536. It has a maximum graphics resolution of 720 by 480 pixels and supports up to 50 rows of text per screen – provided your eyes can stand the strain! It uses the 15 pin connector and analogue signals. Both VGA and MCGA are electrically incompatible with EGA and earlier standards but are software compatible – i.e. software needing CGA or EGA will run correctly but a CGA monitor will not work with a VGA board.

This standard has been effectively superseded by SVGA.

SVGA – Super Virtual Graphics Array

Not content with 'standard' VGA, board manufacturers started to add other high resolution modes. The term *Super VGA* was coined to refer to these extensions. In *Super VGA* mode, resolutions up to 1,024 by 768 are possible – but there are some incompatibilities between the different implementations. As a result, special software drivers are needed to use these modes with Windows and other software, such as Autocad.

The TIGA standard is an interface designed to prevent incompatibilities by defining which modes are supported and how they can be used. Some graphics boards now claim TIGA compatibility, although this standard has not been universally adopted.

EVGA — *Extended Virtual Graphics Array*

This is the most recent graphics standard. It supports up to 256 colours with a resolution of 1,024 by 768 although, at the time of writing, very few systems use it.

The next problem concerns the suitability of the monitor for the graphics board. With earlier designs, this was not a major problem – if you used a CGA compatible board with a CGA monitor, there would be no difficulty. Similarly, using a VGA compatible monitor with a VGA graphics board would pose no problem – provided only the standard modes were used. Unfortunately, most VGA boards support other modes which may not be compatible with a VGA monitor. The reason for this is simple – the additional modes require horizontal and vertical timing which is different from the original specification. Some monitors may require adjustment to work whilst others will not work at all. The proliferation of different timings with VGA is one reason for the popularity of the multisync monitor, which automatically adjusts to different timings.

The different graphics modes supported by standard and enhanced VGA are:

Mode	Screen	Character	Colours	Type
0,1	320 x 200	8 x 8	16	TEXT
0*, 1*	320 x 350	8 x 14	16	TEXT
0+,1+	360 x 400	9 x 16	16	TEXT
2,3	640 x 200	8 x 8	16	TEXT
2*, 3*	640 x 200	8 x 8	16	TEXT
2+,3+	720 x 400	9 x 16	16	TEXT
4,5	320 x 200	8 x 8	4	Graphics
6	640 x 200	8 x 8	2	Graphics
7	720 x 350	9 x 14	Mono	TEXT
7+	720 x 400	9 x 16	Mono	TEXT
D	320 x 200	8 x 8	16	Graphics
E	640 x 200	8 x 8	16	Graphics
F	640 x 350	8 x 14	Mono	Graphics
10	640 x 350	8 x 14	16	Graphics
11	640 x 480	8 x 16	2	Graphics
12	640 x 480	8 x 16	16	Graphics
13	320 x 200	8 x 8	256	Graphics
50	640 x 480	8 x 16	16	TEXT
51	640 x 473	8 x 11	16	TEXT
52	640 x 480	8 x 8	16	TEXT
53	1056 x 350	8 x 14	16	TEXT
54	1056 x 480	8 x 16	16	TEXT
55	1056 x 473	8 x 11	16	TEXT
56	1056 x 480	7 x 8	16	TEXT
57	1188 x 350	9 x 14	16	TEXT
58	1188 x 480	9 x 16	16	TEXT
59	1188 x 473	9 x 11	16	TEXT
5A	1188 x 480	9 x 8	16	TEXT

Mode	Screen	Character	Colours	Type
5B	800 x 600	8 x 8	16	Graphics
5B	800 x 600	8 x 8	16	Graphics
5C	640 x 400	8 x 16	256	Graphics
5D	640 x 400	8 x 16	256	Graphics
5E	800 x 600	8 x 8	256	Graphics
5E	800 x 600	8 x 8	256	Graphics
5E	800 x 600	8 x 8	256	Graphics
5F	1024 x 768	8 x 16	16	Graphics
5F	1024 x 768	8 x 16	16	Graphics
60	1024 x 768	8 x 16	4	Graphics
61	768 x 1024	8 x 16	16	Graphics
62	1024 x 768	8 x 16	256	Graphics
62	1024 x 768	8 x 16	256	Graphics

Note: Modes 0, 1, 2 and 3 are CGA compatible.
Modes 0*, 1*, 2* and 3* are EGA compatible.
Modes 0+, 1+, 2+ and 3+ are VGA compatible.
Modes 0 to 13 are VGA only.
Modes 50 to 62 are extensions.

VGA graphics modes

Mode	Clock MHz	Horizontal sync - kHz	Vertical sync-Hz	Polarity (H, V)
0,1	25.175	31.4	70	+,-
2,3	25.175	31.4	70	+,-
0*, 1*	25.175	31.4	70	-,+
2*, 3*	25.175	31.4	70	-,+
0+,1+	28.322	31.5	70	+,-
2+,3+	25.175	31.4	70	+,-
4,5	25.175	31.4	70	+,-
6	25.175	31.4	70	+,-
7	28.322	31.5	70	+,+
7+	28.322	31.5	70	+,+
D	25.175	31.4	70	+,-
E	25.175	31.4	70	+,-
F	25.175	31.4	70	-,+
10	25.175	31.4	70	-,+
11	25.175	31.4	60	-,-
12	25.175	31.4	60	-,-
13	25.175	31.4	70	+,-
50	25.175	31.5	60	-,-
51	25.175	31.5	60	-,-
52	25.175	31.5	60	-,-
53	40.000	31.2	70	-,+
54	40.000	31.2	60	-,-
55	40.000	31.2	60	-,-
56	40.000	31.2	60	-,-
57	44.900	31.5	70	-,+
58	44.900	31.5	60	-,-
59	44.900	31.5	60	-,-
5A	44.900	31.5	60	-,-
5B	36.000	35.2	56	-,-
5B	50.350	48.0	72	-,-
5C	50.350	31.5	70	-,+
5C	25.175	31.5	70	-,+
5D	50.350	31.5	60	-,-
5D	25.175	31.5	60	-,-
5E	57.272	29.5	90	+,+
5E	25.175	35.2	56	-,-
5E	40.350	48.0	72	+,+

Mode	Clock MHz	Horizontal sync - kHz	Vertical sync-Hz	Polarity (H, V)
5F	44.900	35.2	86	+,+
5F	65.000	48.7	60	+,+
5F	65.000	56.4	70	+,+
60	44.900	35.5	86	+,+
61	44.900	37.9	70	+,+
62	44.900	35.5	86	+,+
62	65.000	48.7	60	+,+
62	75.000	56.4	70	+,+

Note: Modes 0, 1, 2 and 3 are CGA compatible.
Modes 0*, 1*, 2* and 3* are EGA compatible.
Modes 0+, 1+, 2+ and 3+ are VGA compatible.
Modes 0 to 13 are VGA only,
Modes 50 to 62 are extensions.

VGA timings

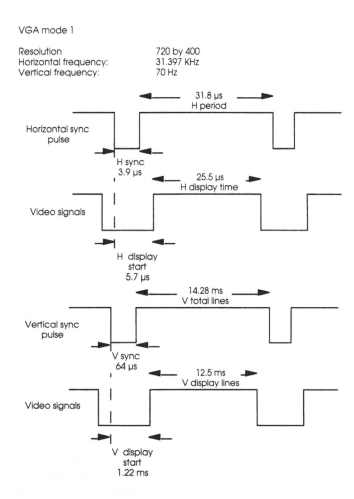

VGA mode 1

Resolution	720 by 400
Horizontal frequency:	31.397 KHz
Vertical frequency:	70 Hz

VGA mode 1 timings

Modes 0, 1, 2 and 3 can have three different sets of values. Mode 0 can be 320 x 200 or 320 x 350 or 360 x 400. The reason for these differences is the display adapter used. If mode 0 is used with a VGA board, the number of colours and characters displayed on the screen is the same as that of a CGA adapter but the character size is increased to compensate for the different screen sizes, so the mode 0 character on a CGA monitor would be 8 pixels by 8, while that on a VGA system would be 9 x 16. The character on such a system running in CGA compatible mode would actually look better because of the larger character size.

Mixing video and graphics

With multimedia applications, the fundamental question about video concerns how the two sources are combined to create a single image. This challenge becomes particularly interesting when the PC is considered which moves objects such as windows and cursors with reckless abandon.

For multimedia systems like CD-i, the solution is to predefine several planes, with each one allocated to a particular source. The planes are ordered hierarchically so that the bottom plane is only seen if the planes above it do not contain any objects. This use of multiple planes and transparent pixels is fine for dedicated systems but unfortunately not when trying to work with an existing video/ graphics system such as a PC.

With PCs, the two most common techniques are both forms of multiplexing, where two or more independent video sources are combined. The basic premise is simple and has already been covered. Essentially, the two video sources are defined as separate planes with a third plane deciding which of the first two planes is used to supply the pixel information to make up the picture. This can be done in two domains: at the analogue level and at the digital level.

Analogue multiplexing

With analogue multiplexing, the video sources — usually just two but the concept can be expanded to more — have the same analogue interface and timings. The H and V sync signals are synchronised so that the pixel information is made available at the same time from both sources. A control plane is then used to identify which source is used and this information is also synchronised with the H and V sync signals.

The two video analogue signals are then fed into a very fast switch which is controlled by the switching signal from the control plane. As all three signals are synchronised and thus have the same H sync and V sync timings, the pixel will have the same locations and correspond exactly to the bit in the control plane. If this bit is set, the switch selects the analogue video signal from one source and, if clear, it selects it from the other. The resulting analogue video signal is then a mixture of both sources, depending on the bits within the control plane. The control plane is effectively masking the video from one source with another. This is thus often referred to as a mask or masking plane.

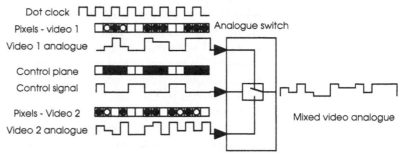

Analogue multiplexing

The main problem with this technique is that the switch has to be at least as fast as the dot clock, which is running at possibly 80 to 100 MHz with a big display, and must have very fast switching times to prevent the analogue signal from being degraded.

With a PC, a further problem is caused by the control and manipulation of the masking plane due to windows and cursor movement. Typically, video is inserted in its own window and this window, like any other in a PC graphical interface, can be overlapped, moved, resized and have a cursor or menus come down over it. To support this, the software which controls the mask plane must be notified every time there is a change within the PC display. When there is, it must change the masking plane so that video is no longer selected where the new object overlaps or lies over its window.

This manipulation, especially for a cursor which is moving many times a second, is extremely processor intensive. Further problems are caused by the ability to dynamically change screen resolutions. The VGA standards support many different screen sizes which, with Windows '95, can be selected dynamically. Every time the screen size changes, the objects are re-arranged — and this must be reflected in the mask plane to ensure that the video is inserted where it should be and not somewhere else.

Digital multiplexing

This is a similar technique, except that the switching is done at the pixel level and not after the pixels have been converted to an analogue signal. Its disadvantage is that access to the graphics controller is needed, while the analogue technique does not need such access and will work with any compatible adapter.

This technique has a control plane that switches the pixel source to insert the video. Many VGA adapters have a feature connector that allows external pixel data to be inserted in mid-stream in such a way. The key problem is in defining and controlling windows and objects. Even with a separate control plane using digital multiplexing, this problem does not go away.

A solution that has been used is chroma or colour keying.

Colour keying

With colour keying, the window to be used to display the video is assigned a unique colour, known as the colour or chroma key, which should only be used by the window as its background. This colour then takes the place of the mask plane. When a pixel is processed with that colour, a pixel from the video source is used instead. The beauty of this system is than when menus, cursors and windows overlap the window, the background colour is lost as it has been replaced by the objects. This immediately stops the video from being inserted where the overlap exists. As a cursor moves across the video window, the video underneath the cursor is automatically removed. The real elegance of this solution is that the maintenance of the background colour key is automatically performed by the PC as part of its normal operation.

There is a downside: the method relies on one value from the colour depth being allocated as the key. If this colour is used elsewhere in the display, the video data could break through where it is not required or the transparent holes may appear. With the large colour depth displays being used with 16 and 24 bit pixels, this is not as big a problem as it was a few years ago. The chances of using the colour key elsewhere are greatly reduced and are not worth worrying about.

Future techniques

With the advent of faster bandwidth buses in PCs, such as PCI, it should be possible to send the video streams directly across the backplane in a pure digital format and combine them at the graphics adapter.

4   Digital picture compression

The ability to compress digital pictures is a fundamental step in providing digital moving pictures. The compression techniques used to compress a colour picture are also the basis of digital video compression. This chapter goes through the basic techniques used to compress a single digital image and explains how the JPEG compression and related algorithms actually work.

Compression techniques

Whenever multimedia is mentioned, it conjures up the vision of TV quality video on a PC — and yet the technology involved in providing this has stretched both the communication and electronic media industries for several years and is likely to continue to do so. The basic problem is essentially one of too much data for too little bandwidth.

The need for compression

The issue of transmitting video is a problem for many delivery systems such as disk drives, CD-ROM, Ethernet and other digital storage mechanisms. The problem is the amount of data that is generated when digitising an analogue picture. While the digitisation process preserves the data better than an analogue storage method (such as a VCR or even a printed page, which can fade or lose data when copied), it does so by increasing the amount of data that is needed to store it.

Consider an A4 sheet of paper with a 10" by 8" printing area. With a 300 dot per inch Laser printer, each square inch would require 90,000 bits of information. The whole page would need $90,000 \times 10 \times 8$ bits or 900,000 bytes to store it. This calculation assumes that the picture is in monochrome. If each pixel was in 24 bit colour, the amount of data needed to store the picture would increase to over 20 Mbytes!

If this is the case, how can personal computers hold such large amounts of data? The reason is the use of compression techniques.

Restricting the choice of data

Most word processors and desk top publishing packages store printed pages not as an image but as text. The image on the page is greatly reduced in terms of what can be displayed and is limited to letters from a character set, different fonts and sizes, and so on. Whilst in terms of producing written documents this is not a major restriction, and results in the reduction of the amount of storage per page from just under one Mbyte to only a few Kbytes, it does pose problems when trying to reproduce a drawing or diagram.

The drawing restriction has been overcome for many requirements by effectively creating the equivalent of a drawing character set to create circles, lines, rectangles and so on. QuickDraw and PostScript define such basic drawing elements and greatly reduce the amount of space needed to store the image.

Like most things in life, nothing is for free. Although these techniques reduce the amount of data needed to store the image, the encoding and decoding of the image are higher. To print a QuickDraw image requires an interpreter that can take the data describing the objects and create the bit map image that is actually printed. This conversion process is also dependent on printer resolution, with a higher resolution image requiring more processing to smooth curves, and so on. It is for this reason that PostScript printers have their own processing systems with CPU performance and memory matching or even exceeding that of the host computer used to create the image. By reducing the amount of storage needed, the amount of processing needed to produce the finished image is greatly increased.

Data compression through coding

The use of symbols to compress data as described above is not the only way of compressing data. An alternative is to look at the data content and change its coding so that it takes less space. For example, consider the number shown. It consists of many different numbers but some are repeated. Where numbers are repeated, they can be coded differently so that the number is given once followed by the number of repeats. To signify that this coding is used, a special character can be used — in this case, • denotes the sequence with the next number used as the repeated number and the second following number used as the number of repeats. In this way, 00 becomes •02 which actually adds an additional character and enlarges the data rather than compressing it. However, 000000 becomes •06 and provides a halving of the data size.

This simple scheme only supports run lengths of up to 9 characters — 10 if 0 is used as a valid number for the repeat value — but it can be expanded if needed. This technique, known as run length encoding or RLE, is frequently used to compress data — especially where there are large amounts of consecutive data with the same value.

```
0011001100220120012000000000002222222222
```

```
•02•12•02•12•02•22012•0012•09•29
```

```
0011001100220120012•09•29
```

RLE encoding

The problem facing video compression is that the information which must be compressed is often not of a format which lends itself to conversion to symbols and often does not provide the long repeating fields of data essential for efficient compression. To achieve the best compression, the data must be manipulated both to reduce the amount of data that must be compressed and to change its format so that encoding techniques are as efficient as possible.

Using the human eye response

One big advantage that the human eye has over a computer screen or electronic camera is its spectral response to colours. Unlike the PC screen, which typically uses the same number of bits for red, green and blue (RGB) images, the eye's sensitivity is such that the weighting is different and, as a result, it becomes difficult to differentiate between colours that the PC is capable of displaying. If the eye cannot discriminate between all the colours, data can be saved by artificially restricting the data that the PC displays. In other words, if the eye cannot discriminate between 16 or 24 bits per pixel, why waste bandwidth transmitting all the data?

This data reduction to match the human eye's response is normally done by converting RGB data into a format which uses a different sampling scheme.

RGB colour space model

As stated before, most computer displays and input devices use the RGB colour space model where red, green and blue pixels are represented by a number of bits per component. With a 24 bit pixel, eight bits are allocated to red, eight to blue and eight to green. With other pixel sizes such as eight or sixteen, the split between RGB is not equal and it is possible to get colour skewing where there may be more blue and red shades compared to green. This skewing is dependent on how the bits are distributed amongst the three colours.

Whilst this representation is easy both to understand and to implement digitally, it does have some drawbacks. Colour changes invariably require changes of all three components and the colour space model does not exactly match the human eye as mentioned previously.

Luma/Croma Representation

The next stage is to convert the RGB pixels into their luminance and chrominance components. Luminance or luma (Y) is the perceived intensity of the image derived from the RGB amplitudes and is calculated by the following equation:

$$Y = 0.299R + 0.587G + 0.114B$$

Chrominance or chroma is the perceived colour of the image and can have three components calculated as follows:

$$R - Y = -1.72 (G - Y) - 0.678 (B - Y)$$
$$B - Y = -2.53 (G - Y) - 1.47 (R - Y)$$
$$G - Y = -0.581 (R - Y) - 0.394 (B - Y)$$

Only the first two chroma components are normally used because the third can be calculated from the other two. These relationships are further processed to define a YUV model where the colour difference signals $(R-Y)$ and $(B-Y)$ may be normalised to form a YUV colour space where:

$$V = (R - Y) / 1.14 = 0.877 (R - Y)$$
$$U = (B - Y) / 2.03 = 0.493 (B - Y)$$

From these values, the CCIR601 recommendation defined a colour model called YCrCb where Y was the luminance component without any change and Cr and Cb were calculated as follows:

$$Cb = 1.144 U$$
$$Cr = 0.813 V$$

YCrCb sampling schemes

With the conversion complete, the picture is represented as a luminance component and two chroma components with the same amount of data per pixel for all three components. Such a colour space is sometimes referred to as a 4:4:4 model.

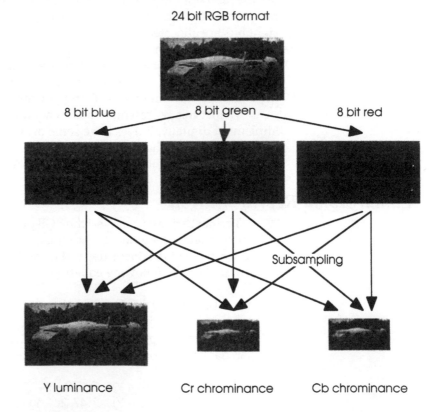

24 bit RGB format

8 bit blue   8 bit green   8 bit red

Subsampling

Y luminance   Cr chrominance   Cb chrominance

4:2:0 colour space conversion

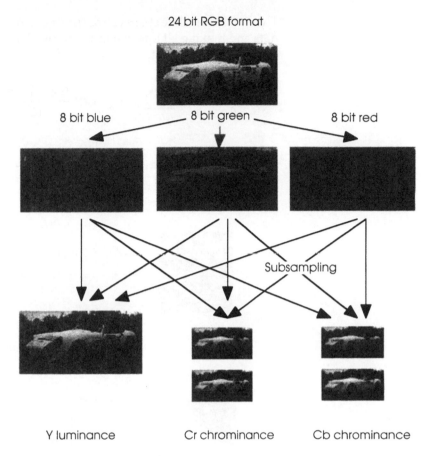

4:2:2 colour space conversion

However, data can be lost by reducing the resolution of the chrominance components by down sampling so that alternative colour models can be used. Models such as 4:2:2 and 4:2:0 are quite common. This is done by averaging out the pixel values in the chrominance components so that a pixel value is shared by multiple luminance pixels.

Using the DCT

One of the most common data transformations used in video compression is the discrete cosine transformation or DCT for short. It is used both in the H.320 group of standards for video conferencing and in both the MPEG1 and MPEG2 standards. Its appeal is based on how it processes data in such a way that it can be more efficiently compressed using RLE techniques. The mathematical equation is quite complex and requires many calculations to be performed.

The equation takes a group of pixels that have been converted into a YCrCb format and processes them into the frequency domain. As a result, the data is redistributed so that the bulk of the information is stored in one corner, with virtually no data left in the rest of the box.

The amount of white space can be further increased by quantising the data and making all low values the same. This makes RLE compression far more efficient.

$$s_{uv} = \frac{1}{4} C_u \, C_v \sum_{x=0}^{7} \sum_{y=0}^{7} S_{yx} \, \cos\frac{((2x+1)u\pi)}{16} \, \cos\frac{((2x+1)v\pi)}{16}$$

where $C_u \, C_v = \dfrac{1}{\sqrt{2}}$ for $u, v = 0$; $C_u \, C_v = 1$, otherwise

The DCT equation

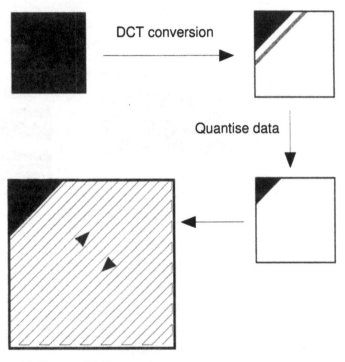

DCT processing

To provide longer sequences of zeros, the encoding is performed in a zig-zag pattern instead of the more normal left to right horizontal method.

JPEG and Motion JPEG

JPEG is a standard that was defined by the Joint Photographic Experts Group to define a standard method of compression for photographic images. The aim was to provide high compression ratios and thus remove one of the problems with high quality digital images.

It is a lossy technique in that data is lost to help increase the compression ratios. The amount of data loss is determined by sets of parameters supplied to the encoder when the image is compressed. In many implementations, the user is presented with a slider bar that determines how lossy the compression is. The program then calculates the appropriate parameters.

It was designed to work with real life colour of grey scale images and exploits many of the techniques already outlined in this chapter. Each pixel is represented using a 24 bit RGB pixel and this provides a wide range of colours. As a result, it is very effective with photographs or computer artwork that have a lot of detail. It is less successful with coping with line art or where there are a lot of sharp edges, as is often the case with cartoons and computer artwork. For a 24 bit/pixel colour image, a typical lossless compression achieves a ratio of about 2:1. By comparison, JPEG can achieve ratios in the region of 10:1 to 20:1 without compromising the visible quality of the image. Although actual data is lost, the image does not appear to lose any detail to the viewer. In other words the data loss is not discernible by the human eye. Increasing the amount of data loss, albeit with some visual degradation, can allow JPEG to obtain compression ratios of up to 100:1. These high ratios are ideal for generating thumbnails for archives while not taking much space and typically having a better visual quality compared to other simpler data loss compression algorithms.

It must be said that JPEG coding is designed for the compression of digital still images, but the techniques it uses have formed the basis of the MPEG specifications which define compression techniques for video. In addition, Motion JPEG is also another standard for encoding video which simply relies on the standard JPEG coding to process each frame in a movie and transmit it as a continuous bitstream. In this case, the frames are independent and cannot be used to help in the encoding/decoding process — a technique that MPEG exploits to obtain higher compression ratios.

JPEG encoding

The encoding process consists of several stages along with options that determine the efficiency of the compression and the amount of data that is lost.

Converting to a different colour space

The first part of the process involves transforming the image into a more suitable colour space for the quantisation/down sampling stage that follows it. With a greyscale image with no colour components, this stage is not performed and the data is further processed as is. With an RGB image, better compression is achieved by translating into a luminance/chrominance colour space such as YCbCr or YUV, where more information can be lost in the chrominance area because the human eye is less sensitive to these high frequency components. This data loss is not performed at this stage but is done at the next operation. Greyscale images are effectively already in this format because they can be thought of as colour pictures with only the luminance part present with no chrominance signal at all.

The algorithm does not insist on the transformation: other compression techniques do not care what the data content actually is but simply work on each component in turn. This means that RGB images can be used without this transformation. However, the quantisation will have a bigger and more visible effect on the image because it will not be masked by the limitations of the human eye.

Down sampling the image

This is an optional process that takes the components from the previous stage and down samples that data to reduce it. This is normally done by averaging together groups of pixels within each component. With a luminance/chrominance image, the luminance component is not down sampled to maintain the resolution. This means that for greyscale pictures with no chrominance components, the image is not down sampled and this stage does nothing.

The chrominance components are down sampled by a factor of two horizontally and either two or one vertically. With a vertical factor of one, there is effectively no down sampling. The two possible schemes are referred to as 2h2v or 2h1v or alternatively, 411 and 422. The data reduction is 50% and 33% respectively.

Down sampling does not work for RGB images where the luminance and chrominance components are distributed between the red, green and blue colour components. While the technique can be used, the resulting image will have visibly degraded and this stage is therefore not recommended for RGB images.

DCT transformation

The next stage is to divide the image into blocks, each block containing 64 (8 × 8) pixels. The block is then transformed into a frequency domain using a discrete cosine transformation. This effectively changes the representation of the data in the block from one that displays the visual image to one where the components are grouped depending on their frequency. As previously described, the human eye is less sensitive to the higher frequency components and these can be removed without impacting visible quality.

Quantisation

Quantisation involves dividing the 64 frequency components by a quantisation coefficient (QC) and rounding the result to the nearest integer. The result is a set of components which are better suited for data encoding. Again, the higher frequency components are normally reduced far more than the low frequency and as a result, the data loss does not reduce the visible image quality due to the human eye sensitivity.

The choice of coefficient is critical to obtaining the best quality of image for a given compression ratio. Most JPEG compression tools simply use the tables specified in the JPEG standard itself or a multiple of their values.

Encoding

This step takes the 8 × 8 block and encodes it using either Huffman encoding or arithmetic encoding. Huffman is almost universally used because arithmetic encoding is optional and only gives about 5 to 10% better compression. The main disadvantage with arithmetic encoding is that it is patented technology and requires a licence to use it. In both cases, the encoding and compression preserve the original data and therefore do not degrade the image any further. At this point the algorithm and compression are complete. All that is left to do is to encapsulate the JPEG bit stream into a file format.

File format

The final part is to add the appropriate file format headers to create a recognisable file which can be interchanged. Generally speaking, there are two types of format: those that include all the parameters so that the decoder can correctly decode the image without any prior knowledge of how the image is encoded and those that do not and assume that the decoder can get this information either from prior knowledge or by having a suitable look up table.

```
                Tuscan test original.JPG best

EOF  =  154413  ($25B2D)                              Data fork

000000:  FFD8 FFE0 0010 4A46   4946 0001 0101 0048   ......JFIF.....H
000010:  0048 0000 FFDB 0084   0001 0101 0101 0101   .H..............
000020:  0101 0102 0201 0202   0302 0202 0202 0303   ................
000030:  0302 0304 0404 0404   0404 0404 0506 0504   ................
000040:  0506 0504 0405 0705   0606 0607 0707 0405   ................
000050:  0708 0707 0806 0707   0701 0202 0202 0202   ................
000060:  0302 0203 0705 0405   0707 0707 0707 0707   ................
000070:  0707 0707 0707 0707   0707 0707 0707 0707   ................
000080:  0707 0707 0707 0707   0707 0707 0707 0707   ................
000090:  0707 0707 0707 0707   0707 FFC4 01A2 0000   ................
0000A0:  0105 0101 0101 0101   0000 0000 0000 0000   ................
0000B0:  0102 0304 0506 0708   090A 0B01 0003 0101   ................
0000C0:  0101 0101 0101 0100   0000 0000 0001 0203   ................
0000D0:  0405 0607 0809 0A0B   1000 0201 0303 0204   ................
0000E0:  0305 0504 0400 0001   7D01 0203 0004 1105   .........}......
0000F0:  1221 3141 0613 5161   0722 7114 3281 91A1   .!1A..Qa."q.2...
000100:  0823 42B1 C115 52D1   F024 3362 7282 090A   .#B...R..$3br...
000110:  1617 1819 1A25 2627   2829 2A34 3536 3738   .....%&'()*45678
000120:  393A 4344 4546 4748   494A 5354 5556 5758   9:CDEFGHIJSTUVWX
000130:  595A 6364 6566 6768   696A 7374 7576 7778   YZcdefghijstuvwx
000140:  797A 8384 8586 8788   898A 9293 9495 9697   yz..............
000150:  9899 9AA2 A3A4 A5A6   A7A8 A9AA B2B3 B4B5   ................
```

JFIF file hexadecimal dump

In the PC world, two types of format are extensively used: JPEG File Interchange Format (JFIF) has established itself as a de facto format (especially on the Internet) and the Apple Macintosh PICT JPEG format, which is essentially the JFIF format coated with a PICT structure. These two formats are very similar and there are many utilities available to convert between them.

There are other formats available, such as TIFF 6.0 and HSI from Handmade Software and Image Alchemy, although these are almost considered proprietary and therefore their appeal is limited.

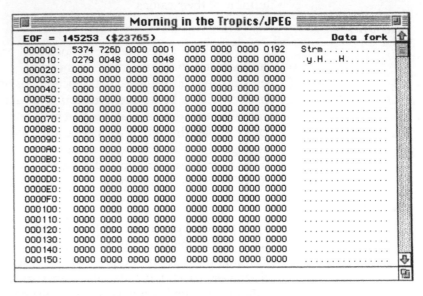

*Storm Technology proprietary
JPEG file format*

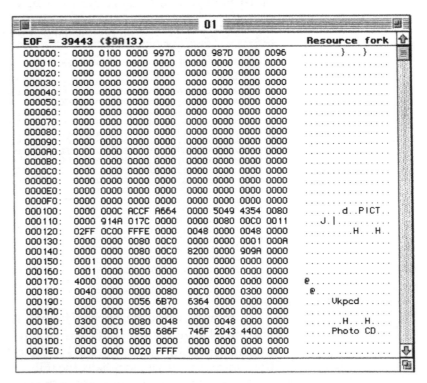

PhotoCD format (PICT based)

File format	Structure
JFIF	Starts with the four bytes (hex) FF D8 FF E0, followed by two variable bytes (often hex 00 10), followed by 'JFIF'.
Raw JPEG	Starts with the two bytes FF D8 and often (but not always) decoded by a JFIF compatible decoder.
HSI	Starts with the text 'hsi1'. Requires HSI compatible software to decode it.
Storm	Starts with the text 'strm'. Requires Storm Technology software to decode it.
PICT-JPEG	Starts with an Apple Macintosh PICT header (typically 726 bytes) that often contains the text 'Photo - JPEG' and 'JFIF' or 'AppleMark'. The actual JPEG data starts with the three byte sequence FF D8 FF. Removing all the header before this three byte sequence often allows a JFIF compatible decoder to use the file.
PhotoCD	Similar to the Apple Macintosh PICT-JPEG format but contains the text ' Photo CD'. These files are normally opened using the slide viewer program stored on the Photo CD itself.

Optional JPEG modes

The JPEG standard describes several different modes of operation although most of them are not used or have not been generally implemented.

The progressive mode

This mode allows the full quality picture to be decoded by a series of multiple scans, with each scan improving the image quality of its predecessor. In this way, a low quality 'preview' can be displayed to allow the user the option of continuing or aborting the complete decode process. The technique is also useful for the real time transfer of images where the low quality preview is only improved if time allows. This method is quite compute intensive in that each image still requires a complete JPEG decode per scan.

The hierarchical mode

With this mode, the image is encoded using different resolutions, where the larger resolution image is encoded as the difference between it and the smaller image. This technique would allow many different image resolutions to be more efficiently compressed and allow the viewer the option of displaying the most appropriate resolution for the system. The disadvantage is that the file size is considerably bigger than that of a single resolution and therefore is only advantageous if the recipient actually needs all the different resolutions.

Lossless JPEG

The DCT operation is not lossless because of rounding errors within the computation. This followed by the quantisation process makes the JPEG encoding process lossy, although the visual impact of this data loss is kept to an absolute minimum. The lossless mode uses the basic JPEG encoding algorithm but does not lose data. The compression ratios are obviously not as good but all the data is preserved.

To achieve this, the lossless mode does not use colour space conversion, down sampling, DCT or quantisation but simply codes the difference between each pixel and the predicted value for the pixel. The Huffman or arithmetic encoding is still used to provide additional data compression.

Even with the lossless mode, data can be lost or at worst misinterpreted if the encoder and decoder handle errors and rounding in different ways.

JPEG decoding

The decoding process for a JPEG file is the inverse of the decoding process. The bitstream is decoded and the coefficients extracted along with the quantisation information that was used to create them. An inverse quantisation is then performed to reconstruct the results of the DCT. An inverse DCT is then performed to create the macroblocks and thus create the picture output.

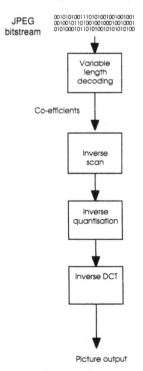

*The JPEG decoding process
block diagram*

Re-encoding JPEG files

As the JPEG algorithm uses a lossy compression technique, artefacts can be created in an image, especially if the parameters are set for maximum compression. The set of pictures overleaf shows the effect of JPEG compression using the highest compression and highest quality settings of a JPEG utility. Each picture includes information of the original file type, its size and the time to decompress the file. The pictures were taken from an Apple PowerMAC 7100.

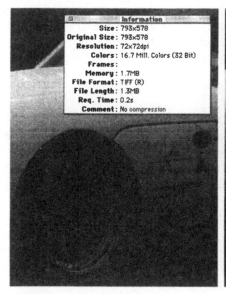
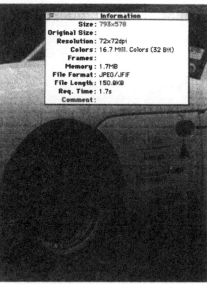

Original TIFF format file and JPEG encoded with highest quality

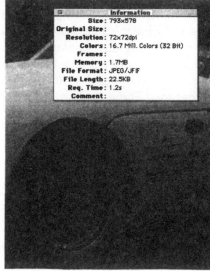
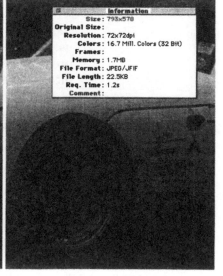

JPEG encoded with lowest quality (left) and JPEG re-encoded with lowest quality (right)

The lowest quality JPEG picture introduces artefacts into the picture: the lettering to the right of the car wheel is blurred and marks appear either side of the crossed line motif. In general, the image quality where the picture goes from white to black is also downgraded.

It is possible to re-encode an already JPEG encoded picture. This is generally not a good idea because the image is greatly degraded and the compression achieved is often zero. The reason is that both the initial and last compression used the same parameters and therefore achieve the same result. In practice, repeated JPEG compression simply continues to downgrade the picture. If higher compression is needed, it is better to compress the original picture with different JPEG parameters.

JBIG

JPEG is a lossy compression technology which does not work well with black and white 1 bit images or where there is sharp contrast between objects. To address these limitations, another group of experts was formed called the Joint Bi-level Image Experts Group or JBIG. Its main goal was to define a lossless compression algorithm which supported both sequential and progressive decoding.

A progressive decoder is one that can create a lower resolution picture from the bit stream, without having to process all the information. It can, if needed, fill in the detail later — but the main advantage that progressive decoding provides is the ability of the encoder to only perform the amount of work needed to extract the image to the required resolution that the encoder can use. For example, a large A0 size engineering drawing can be progressively encoded using 300 dots per inch resolution. It can be progressively decoded so that only 72 dpi resolution image is extracted to be shown on a workstation screen. This is far more efficient than decoding the whole picture at 300 dpi and then reducing the data down to the lower resolution as it uses less memory and processing power and is thus a lot quicker. The progressive/sequential compatibility is achieved by encoding the picture in strips. The chosen resolution is determined by how much of the strip is decoded.The encoding technology is based around IBM's patented Q coder technology. The encoder does not use multiplies unlike many others that use the multiply accumulate calculation to perform DCTs and so on.

5   Digital video compression

In the last few years, the ability to provide digital video from a CD-ROM, a telephone line or from a satellite or terrestrial broadcast has moved from science fiction to science fact. The technology to provide the compression rates needed has become available and thus the ability to provide good quality digital video has hit the market place.

Digital video has several advantages over its analogue counterpart:

- It is easy to copy and reproduce without losing or degrading quality. This may or may not be a good thing depending on the point of view. It certainly means that the degradation seen with analogue video recording, such as VCR, will not occur — but it does make it easier for pirates to duplicate copyrighted material. As a result, it is not uncommon for encryption to be incorporated into digital video technology to prevent unlicensed duplication.
- It is easy to manipulate in digital format and thus re-use the images. For example, morphing requires the video to be in a digital format before processing.
- It requires less bandwidth to deliver the data. This is becoming increasingly important for radio and satellite based transmissions where the RF bandwidth needed to transmit a single analogue TV channel can support multiple digital channels using digital video compression technology.
- It can incorporate additional data such as subtitles, bibliographies and alternative language soundtracks.
  Digital video does, however, have several disadvantages:
- It requires a lot of processing power to encode and decode the video.
- The equipment and technology for encoding are different from that needed to decode. (For details, see later in this chapter and Chapter 12.)
- Much of the video material available today is still in the analogue domain, yet to be converted.

Video compression techniques

The previous chapter described the basic methods used by the JPEG standard to encode still pictures. It is possible to use these basic standards and treat video simply as a succession of still pictures. This is the idea behind Motion JPEG — but it does not provide sufficient compression or a method of encoding the audio that needs to go with the video to create the complete movie. In fact, in most cases, the term digital video implies the inclusion of audio as well. To improve the compression obtained from the colour space conversion, using the

DCT with quantisation and Huffman encoding, the motion within the frame is analysed and used to derive additional information which can be used to reconstruct the picture.

Sending the differences

If you look at the individual frames in a movie, there is often very little difference between adjacent pictures' content. It is this small difference, coupled with our persistence of vision, that creates the illusion of motion.

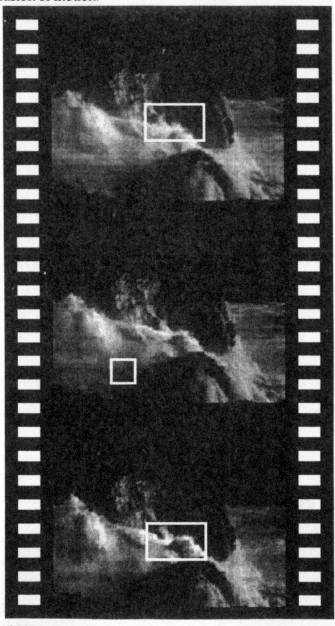

Movie frames — main differences highlighted by white boxes

The three frames shown as an example of a wave breaking on a rock show how small these differences can be. The differences typically take the form of new objects appearing or disappearing in the field of view or small changes to existing parts of the image. The second frame has an object that appears only in the second frame and the highlighted areas on the first and last frame show the main scope of changes within the picture.

It is possible to create the second and third frames simply by transmitting the differences. The second frame can be constructed from the first by adding the object. The third frame can be created by sending the changes to the first frame to create the new central section. In both cases, the amount of data sent is greatly reduced.

Motion estimation

Another technique that can be used with video to dramatically reduce the amount of information needed is to use some form of motion estimation. In its simplest form, motion estimation analyses the video frames and calculates where objects are moved to. Instead of transmitting all the data needed to represent the new frame, only the information (i.e. the vector or new position) needed to move the object is transmitted.

Near fit vector comparison

The difficulty with motion estimation is in defining an object. It can be quite obvious, when there is a high degree of contrast such as where a dark object is moving across a light background. Where such contrast is not apparent, the processing needed to identify an object is very high. There are also complications when objects break or join together. A car may come into the picture and a passenger get out. The passenger is undoubtedly part of the car whilst inside the vehicle. However, when the passenger opens the door and starts to climb out, it could be argued that the car object is simply changing as the passenger is still attached to the car and forms part of the car's image. How can it be judged when the passenger leaves the car and becomes a separate entity?

One method of carrying out motion estimation is to divide the picture into arbitrary blocks instead of trying to identify objects and compare the block with its neighbours. The resulting differences can be encoded to allow the decoder to build up the picture.

MPEG1 overview

The MPEG1 standard for audio and video compression is an international open standard developed by the Motion Picture Experts Group to provide a compression method for high quality audio and video. It was designed to complement the then emerging CD-ROM technology and typically uses a bandwidth of 1.2 Mbits per second — the data rate obtained from a single speed CD-ROM player. It relies on more sophisticated compression techniques to support larger picture sizes and provide better quality images. The aim was to provide good quality video and audio and rely on hardware acceleration to perform the processing or the ever more performant PC to have sufficient processing to perform the task in software.

Like Indeo (the Intel proprietary video and audio standard), it uses colour sub-sampling but at a 25% reduction level instead of Indeo's 6.25%. However, it does not use the lossy vector quantisation technique but motion estimation and discrete cosine transform techniques to improve its quality, especially when used with fast moving subjects and images.

One of the problems associated with extracting frame differences is in identifying where the changes have occurred. The MPEG1 standard defines a motion estimation algorithm which efficiently searches multiple blocks of pixels within a given search area and thus easily tracks objects which move across the screen or within the frame as the camera pans around.

The second technique that MPEG1 uses is discrete cosine transform (DCT). This processing is performed after the colour sub-sampling and again exploits the physiology of the human eye. It takes a block of pixels and converts them from the spatial domain to the frequency domain. This conversion allows the higher frequency components to be reduced as the human eye is less sensitive to these components and it improves the efficiency of the RLE compression. As a result of these methods, the picture quality is far superior to that of Indeo compressed video.

MPEG1 video compression

MPEG1 video compression (the audio compression techniques are covered later in this chapter) typically compresses a SIF picture of 352 by 240 pixels at 30 frames per second (US NTSC format) or 352 by 288 pixels at 25 frames (Europe PAL format) per second as its basic input format. Although formats which use larger pixel frames (such as CCIR-601 with a pixel format of 704 x 480) are supported, the SIF formats are the most frequently used. SIF formats are actually derived from the CCIR-601 format due to interlacing effects. The CCIR-601 digital television standard, which is used by professional digital video equipment, defines a picture size of 720 by 243 or 240 by 60 fields (not frames) per second. A frame actually comprises two fields, where the odd and even information is interlaced to create the full picture. While the interlaced luminance information occupies the full 720 by 480 frame, the chrominance components are reduced by down sampling so that they contain less data and are 360 by 243 or 240 by 60 fields a second. In effect, each chrominance pixel is shared between several luminance pixels. These formats both use 4:2:2 encoding for each pixel.

The MPEG1 standard further downsamples the chrominance components by reducing the pixel data by half in the vertical, horizontal and time directions. It also cuts lines to ensure that the pixel parameters are divisible by 8 or 16. This is necessary because the motion analysis and DCT conversion are performed on 16 by 16 or 8 by 8 blocks. As a result, the number of lines change for an MPEG1 encoded movie between the NTSC and PAL/SECAM standards. The final figures of 288 at 50 fields and 240 at 60 fields per second have the same number of bits and thus have the same common bitstream bandwidth requirements.

The resulting data bitstream is designed for a bandwidth of 1.5 Mbits per second, which is the data transfer rate obtained from a single speed CD-ROM drive. It is no coincidence that MPEG1 is used as the video compression standard for CDi based systems. The actual specification defines a bit rate of between 40Kbit/s and 1.86Mbit/s. The bitstream itself comprises three components: the compressed video bitstream, the compressed audio bitstream and a system level. Both the audio and video streams are time stamped to allow easier synchronisation and provide good lip synching.

The constrained parameter set

Like the JPEG standard, the MPEG1 standard allows the user to set up a whole range of parameters which control the image size, target bitstream rate, and so on. However, to provide a level of commonality, a constrained parameter set has been defined and most MPEG1 decoders conform to this. This also means that for parameters outside and beyond the constrained set, there is less likelihood of compatibility between different encoders and decoders.

- Horizontal resolution ≤ 768 pixels
- Vertical resolution ≤ 576 pixels
- Macroblocks per picture ≤ 396
- Macroblocks to be processed per second ≤ 99000
- Frames per second ≤ 30
- Bitstream bandwidth ≤ 1.86 Mbps
- Decoder buffer size ≤ 376832 pixels

MPEG1 video compression process

Colour space conversion

The first stage is to convert the video picture into its correct colour space format. This is an optional stage depending on the originating format. In most cases, the incoming data is in the RGB format and thus will need converting.

The format used is 4:2:0 YCrCb and this involves subsampling and some data loss and hence compression. For more details on the colour space conversion, refer to the relevant sections in the previous chapter.

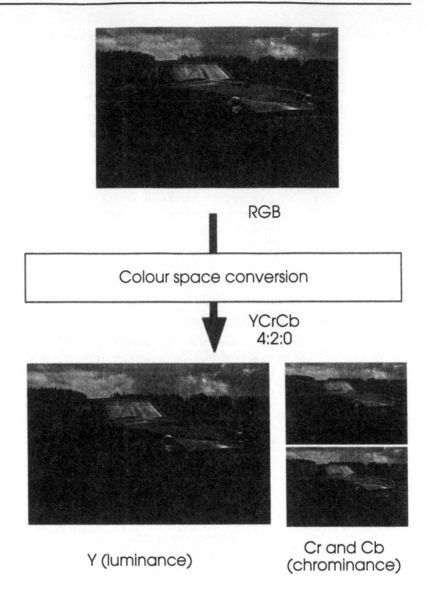

RGB

Colour space conversion

YCrCb
4:2:0

Y (luminance)

Cr and Cb
(chrominance)

Colour conversion

Slices and macroblocks

The motion estimation process within MPEG1 is based on the concept of macroblocks. Each frame is divided into a set of slices, each slice containing eleven macroblocks. There are different numbers of slices and hence macroblocks, depending on the video source. The diagram is for a PAL picture, which has a 352 by 288 pixel frame — 36 slices each containing 11 macroblocks. This gives a total of 396 macroblocks for the total frame.

Each macroblock varies in size depending on the colour component: for the luminance (Y) values, the macroblock is 16 by 16 pixels. For the two chrominance components, the macroblock is 8 by 8 pixels. Only the luminance component is used for the motion estimation calculations.

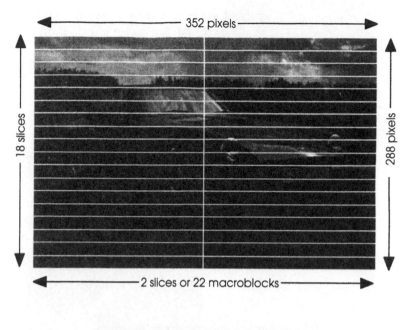

1 slice = 11 macroblocks

1 macroblock = 16 by 16 pixels (Y) or
8 by 8 pixels (Cr or Cb)

Slices and macroblocks

Motion estimation

The basic principles behind the motion estimation algorithms used in MPEG1 are very simple. Each macroblock is compared with other macroblocks within either a previous or future frame to find a close match. The comparison is performed on the 16 by 16 pixel macroblock containing the luminance (Y) data. If there is close match, the information can be used instead of the original data. When a close match is found, a vector is sent to describe where this block should be located and any difference information for this block. This is far more efficient than sending either the original or matched data.

The diagram shows how this can be done: the macroblock to be compared is located at a vector reference of 0,0. The macroblock for comparison is selected from another frame — either a previous or future frame can be used — and the pixels compared mathematically. If they are not the same or a close match, the search continues by selecting another macroblock at a slightly different offset. The offset can be either on a macroblock boundary or on a pixel boundary. The comparison is then repeated until a match is obtained or the specified search area within the frame has been exhausted. If no match is available, the search process can be repeated using a different frame or the macroblock can be sent as a complete set of data and this is then passed to the DCT converter for conversion prior to encoding. If a match is found, the vector information specifying where the matching macroblock can be located is used instead, together with any difference information that can be used to improve the accuracy of the match. This information is again sent to the DCT converter prior to encoding.

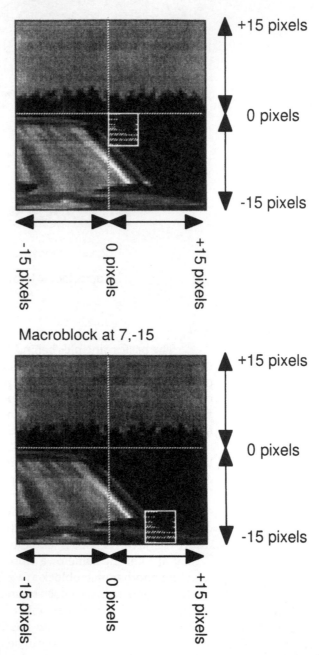

Macroblock at 0,0

Macroblock at 7,-15

Motion estimation and search area

This technique is extremely compute intensive due to the number of comparisons that need to be made, the size of the search area, the number of macroblocks that must be processed and the frame rate. This in itself has several implications:

- The compression process becomes asymmetrical, with more processing power needed to encode the compressed bit stream than to decode it. Typically, encoding is done in non-real time by very powerful workstations. This limits the use of MPEG1 material to effectively pre-recorded material, although silicon chips are now appearing which reduce the encoding time.
- The encoder now influences the quality of the decoded image dramatically. Encoding shortcuts, such as limited search areas and macroblock matching, can generate poor picture quality — irrespective of the quality of the decoder.
- The decoder needs a large amount of memory to store the previous and future frames needed to allow the motion estimation information to be decoded.

With the motion estimation completed, the raw data describing the frame can now be converted using the DCT algorithm ready for Huffman encoding.

DCT conversion

The principle behind DCT encoding was described in the previous chapter; it is the same as used in JPEG still image conversion. The macroblock information for the luminance and chrominance components is converted using the discrete cosine transformation into a frequency domain. The data is divided or quantised to remove the higher frequency components and to make more of the values zero. This is depicted by the increased white area in the square. This is an extremely compute intensive process.

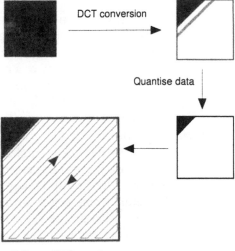

DCT conversion

Quantise data

Huffmann/RLE encode

DCT processing

Encoding

The next stage is to encode the data. All the compressed data is Huffman encoded with a set of fixed tables to create the video bitstream. The DCT co-efficients are encoded using a variation on the normal encoding method, where the Huffman code not only specifies the number of zeros but also the value that ended the run of zeros. This value is obviously a non-zero value. This is more efficient, especially as the zig-zag encoding method is used.

I, P and B frames

One of the more complex parts of the MPEG video compression algorithm is its use of different frame types and their subsequent involvement with the motion estimation processing. The MPEG1 standard defines three types of frame.

The I or intra frame

This is the starting point for the whole process. An intra or I frame is one which is treated as a still process and has simply been DCT processed but does not use or require any prior knowledge of past or future frames to decode it. In other words, the motion estimation processing has not been performed on this type of frame. It has several uses: it provides a known starting point and is usually the first frame to be sent.

I frames are often used as internal references for trick modes, such as the ability to fast forward through a picture. MPEG1 does not support these types of mode and they have to be simulated. One method is to simply decode and display I frames on the screen. This would involve parsing the bitstream to identify the frames and ignoring any intermediate B and P frames.

The P or predicted frame

This frame uses the preceding I frame as its reference and the motion estimation processing. Each macroblock in the frame is supplied either as a vector and difference with reference to the I frame or, if no match was found as a complete encoded macroblock where all the information is sent, as is the case with the I frame. This is called an intra-coded macroblock. The decoder must retain the I frame information to allow this frame to be decoded.

The B or bi-directional frame

The B frame is similar to the P frame except that its reference frames are the nearest preceding I or P frame and the next future I or P frame. When compressing the data, the motion estimation works on the future frame first, followed by the past frame. If this does not give a good match, an average of the two frames is used. If all else fails, the macroblock can be intra-coded.

Needless to say, decoding B frames requires that many I and P frames are retained in memory.

MPEG1 *frame sequence*

With the interdependency between the different frame types, the MPEG1 frame sequences are a little confusing, to say the least. Frames need not be in order, although a large amount of re-ordering is very inefficient and requires a large number of frame buffers to hold the frames until all the dependencies have been cleared.

The main framework is based around the need for an I frame on a regular basis to enable some form of random access to be supported and also allow trick modes, such as fast forward, to be supported. Typically, this means sending an I frame every 0.4 seconds which works out as 12 frames between each I frame, assuming a 30 frame per second rate is used, as is the case in the US and Japan but not elsewhere.

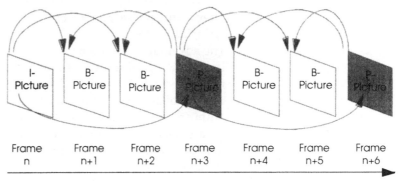

MPEG1 frame sequence

With a rate of 30 frames per second, the sequence consists of a starting I frame followed by two B frames, a P frame followed by two B frames, and so on. This sequence is sometimes referred to as a group of pictures or GOP.

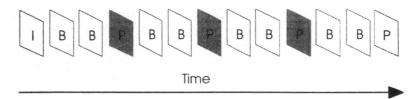

MPEG 1 GOP structure

For a decoder, this sequence is interesting in that the I frame is first received, stored and decoded. The next two B frames have to be stored locally until the P frame arrives. The P frame can be decoded using the stored I frame and the two B frames can be decoded using the I and P frames. One solution to this is to reorder the frames so that the I and P frames are sent together followed by the two intermediate B frames. Another more radical solution is not to send B frames at all and simply use I and P frames.

The inclusion of B frames within the MPEG1 standard is one area where there is not total agreement within the industry, although many objections raised to their inclusion have disappeared with the advent of higher density memories and the increase in processing power. Objections to B frames include:

- Increased computational load during the encode process. The inclusion of the previous and future I and P frames as well as the arithmetic average greatly increases the processing needed. This argument is still valid but with the increased processing power now available, it is less relevant.
- Increased frame buffers to store frames to allow the encode and decode processes to proceed. This argument is again less valid with the advent of larger and higher density memories. However, it is still a sticking point for consumer applications where cost of components such as memory is always critical.
- They do not provide a direct reference in the same way that an I or P frame does.
  The advantage of B frames is:
- Improved signal to noise ratios due to the averaging out of macroblocks between I and P frames. The ability to use an average of these frames means that noise (i.e. random artefacts within the picture) are effectively masked out or reduced. This is particularly useful for lower bit rate MPEG1 applications but is less of a benefit with higher rates, which have improved signal to noise ratios.

Decoding delays

The use of predictive frames can cause one other problem, although for most MPEG1 applications this is not an issue. Whenever an MPEG1 decode is started, there is a delay until all the relevant frames have been received before they can be decoded. This delay is typically 100 to 200 ms and is determined by both the length of time taken to fill the buffers and process the frames and the decoder design itself. Once decoding has started, the delay is not noticed. In reality, the viewer sees the decoded video and hears the audio in synchronisation but some hundred or so milliseconds after the information has been transmitted. This can cause problems when MPEG streams are decoded or used in interactive applications.

MPEG1 audio compression overview

Until now, description of the MPEG1 compression process has concentrated on the compression of the video data. This section now considers how the MPEG1 audio is compressed.

The aim of the MPEG1 committee in defining the audio compression techniques it would specify, was to provide CD quality for the listener. This in itself was quite a tall task without compression because the raw data rate for CD audio is about the same as that for the single speed CD-ROM. For MPEG, the CD-ROM data rate of 1.5 Mbits per second must be shared between data and video.

With an audio CD, the sampling rate is 44.1 kHz and the sample size is 16 bits to provide the dynamic range and signal to noise ratio that is required. The audio is in stereo and therefore the raw data rate is $16 \times 2 \times 44{,}100$ bits per second, which equals 1,411,200 bits per second — just under the 1.5 Mbits per second CD data rate. MPEG1 video is typically allocated about 1.25 Mbits per second for video and this leaves about 250 kbits per second of bandwidth for audio. To fit into this allocation, audio compression of about 6:1 is needed. The problem that must be solved is how to compress the audio without compromising the audio quality heard by the listener.

Audio compression options

To reduce the audio bit stream, three fundamental methods can be used:

- Reduce the sampling rate so that there is less data to send.
  This sounds fine, except that the laws of physics get in the way! Nyquist's theorem states that the sample rate must be at least twice the frequency of highest sampled frequency that must be reproduced to prevent artefacts from being introduced into the audio. This is the reason that CD audio sample rates are 44.1 kHz, which gives a maximum frequency of 22.05 kHz — comfortably above the 20 kHz frequency that most audio systems support. Reducing the sample rate reduces the number of bits needed — but it is not transparent to the listener.

- Reduce the sample size so that there is less data.
  With this approach, the sample size is reduced from 16 bits to a smaller figure, such as 8 or less. This again is not a transparent compression technique. The 16 bit sample size was chosen for several reasons: it provides a large signal to noise ratio of about 90dB, which is a good match for the human ear. This means that noise generated during the sampling process due to quantisation effects is not heard by the listener. This is the reason that CD audio sounds better than analogue audio from a vinyl record: quiet passages are quiet and loud passages are loud. Reducing the sample size destroys the signal to noise ratio and thus noise would be heard during quiet passages.

- Use coding techniques such as Huffman to reduce the amount of data that is sent.
  This is a good technique which preserves data but unfortunately does not achieve the compression ratios that are needed and therefore it must be augmented by other techniques.

Using the psycho-acoustic model

The solution adopted by the MPEG1 committee attacked the problem on two fronts: it reduced the data rate by reducing the sample size and not the sampling frequency and used Huffman encoding to further compress the audio data.

To reduce the sample data, it used the psycho-acoustic model, which exploits the characteristics of the human ear. In the same way that video compression techniques exploit the eye's lack of sensitivity in the higher frequency video components, allowing them to be discarded without a visible loss of video quality, the psycho-acoustic model allows certain data samples to be reduced in size without affecting the audio quality heard by the listener.

The masking effect

The masking effect is a phenomenon where noise is only heard when there are no other sounds to mask it. For example, the high frequency hiss from an amplifier or rumble from a turntable is only heard during quiet passages. If there is music playing, the louder music masks out the quieter rumble and hiss and they are not heard. This has been exploited by unscrupulous audio sales personnel for a long time: the customer hears the system in the shop playing music that has little or no quiet passages and therefore does not hear any hiss or noise. When the customer plays a classical piece of music with quiet passages at home the noise is no longer masked and is thus audible. In reality, the noise component is always there, but the audio processing that our brains perform mask out the noise so that we only hear the sounds we want to hear. When there is nothing to mask the sound, we hear the noise.

This effect also appears when considering sounds and not noise. A very loud sound at a certain frequency will mask out a quieter sound at a similar frequency. As a result the sound heard by the listener will appear to only contain the loud sounds and the quieter sounds will be masked out i.e. reduced in amplitude so that they are either not heard or heard at a greatly reduce volume. The diagram shows the effect in operation.

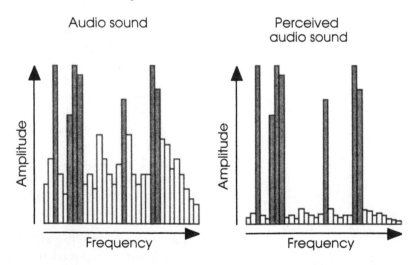

The audio masking effect

Graphic equalisers exploit or counter this effect by boosting the amplitude of specific frequencies and thus countering the masking effect. In this way, specific instruments can be 'brought out' from the sound. The resolution is obviously dependent on how narrow each frequency band is.

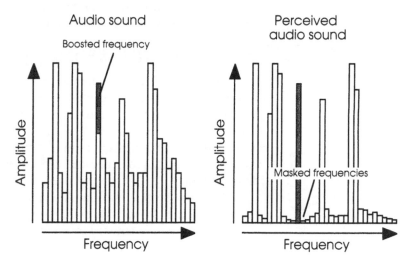

Countering the masking effect

The diagram shows how this can be done. One of the middle frequencies is boosted by the graphic equaliser (depicted by the grey boost) and this appears in the perceived sound. To the listener, it would have appeared as if by magic. The boosted frequency also has its own masking effect and the frequencies around it are also further masked, as shown by their further reduced amplitude, when compared with the previous diagram. Masking effects are additive — another useful advantage for compression.

Given this effect, how can it be used to compress the audio data? If a frequency is going to be masked by other louder frequencies, the noise associated with that frequency is also going to be masked. Therefore, the sample size can be reduced for that frequency without impacting the quality. The reduced sample size means that less data is sent and therefore compression has been achieved. The reduced sample size increases the noise present with that frequency, but it is masked out. This is the basic principle for the MPEG1 audio compression.

Non-linear sensitivities

The psycho-acoustic model also provides some additional help: the ear is not linearly sensitive to all frequencies and, at the extremes of the frequency range (i.e. high and low frequencies), its sensitivity is reduced. Peak sensitivity is between 2-4 kHz, which is the frequency range of the human voice and the bandwidth used by the telephone system (3.1 kHz). Noise in the less sensitive frequency ranges is more easily masked, while it is important to reduce any noise in the peak range to a minimum due to its greater impact.

Pre and post-masking

The masking effect can also occur before and after a strong sound where the amplitude changes by 30 to 40 dB for example. Pre-masking occurs for about 2 to 5 ms before the sound is perceived by the listener and the post-masking effect lasts for about 100 ms after the sound. As a result, there is not only a masking effect taking place across the frequencies but also in time.

The MPEG1 audio compression process

The MPEG1 standard defines three audio compression standards called Audio Level I, II and III. Three levels were chosen because the acoustic tests did not come out in favour of any one approach as all of them were felt not to be completely 100% transparent. In practice, they are transparent as most listeners cannot recognise the artefacts unless they are pointed out or the listener is trained to identify them.

The first level is the simplest and uses the psycho-acoustic model to mask and thus reduce the sample sizes. The Level II coder enhances the Level 1 standard by improving the accuracy and using more advanced techniques. The Level III coder is the most advanced and uses special filter banks and non-uniform quantisation techniques. The three levels are in fact supersets of each other with Level III decoders capable of decoding all three levels and Level II decoders are assumed to be capable of decoding a Level I stream as well as Level II. The highest compression and the most compute intensive coding is Level III, while Level I is the simplest to implement but has the lowest compression ratios. Level II is the most frequently used of the three standards.

Three sample rates are defined which are common to all three levels: 48 kHz for use in professional sound equipment, 44.1 kHz as used in CD-audio and 32 kHz for communications equipment.

The standard supports two audio channels, which can either be single mono channels, a dual mono channel or a stereo channel using either the normal left-right encoding, middle/side (MS) or intensity encoding.

With the MS stereo format, one channel carries the summation of the left and right channels, while the other carries the difference between the two signals. The original left and right channels can be reconstructed by adding and subtracting the difference channel from the summation channel. The original amplitude can be derived by dividing the results by 2.

```
Sum    = l + r
Diff   = l - r
Left   = x + y  = l + r + l - r = 2l
Right  = x -  y = l + r - l + r = 2r
```

Decoding an MS audio stream

With the intensity format, the high frequency parts of the left and right signals above 2 kHz are combined. The stereo image is reconstructed from this data along the temporal envelope i.e. the information that determines the relative amplitude of the left and right channels, as shown in the diagram. In simple terms, the amplitude data for each channel is superimposed on the combined frequency spectrum of the left and right channels. This then reconstructs the left and right channels. This is another application of the psycho-acoustic model, where stereo perception is more a function of the overall audio signal rather than the components within it. In addition, MPEG allows for pre-emphasis, copyright marks and original / copy marks.

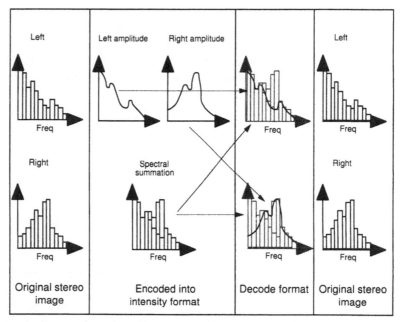

Generating the stereo image
from the intensity format

Both the MS and intensity encoding provide further help in compressing the audio stream.

Level I

The Level I coder is aimed at bit rates above 128 kbits per second per channel and is used, for example, by Philips with the Digital Compact Cassette (DCC) which supports a bit rate of 192 kbits per second. The coder transforms the audio signal into a frequency domain and divides that domain into 32 sub-bands, where each band is 625 Hz wide. This allows each sub-band to be processed using a slightly different acoustic model and thus achieve a better and more accurate encoding. In reality, the decision to have a uniform width for each sub band is a bit of a compromise because the human ear is not linear in terms of its masking effect, especially with higher frequencies. However, the computational load to create non-uniform sub-bands is great and it was decided to adopt a simpler situation. The sub-bands are typically created using a polyphase filter bank, which is relatively fast. However, it does lose some data due to quantisation and it does not provide a clean sub-band — a frequency component at the edge of one sub-band can also appear in its neighbour.

With this transformation completed, the information is processed using the psycho-acoustic model to capitalise on the masking effect. There are two acoustic models used known as models 1 and 2. Model 1 is less complex and used for Level I and Level II encoding while model 2 is used for Level 3.

Model 1 uses a 512 sample window i.e. it processes 512 samples at a time and uses that block of data for the analysis. As a result, the audio is analysed and processed in separate blocks. With level 1 encoding, 32 samples are taken for each of the 32 sub-bands giving a frame of 384 samples which is well within the 512 sample window specified by the psycho-acoustic model 1. With Level 1, the 384 samples are centred by calculating an offset into the middle of the 512 sample frame.

The processing for Level I using model 1 consists of the following stages:

- Convert the audio into the frequency domain.
  This is where the audio is converted from the time domain into the frequency domain using a function such as a fast Fourier transform.
- Extract the critical band widths from the spectral information.
  To simplify the calculations, the spectral information is divided into 32 sub-bands, each with a width of 625 Hz. This operation is often performed with the preceding stage using a polyphase filter.
- Extract the tonal (audio) information from the noise.
  Model 1 treats tonal information (i.e. the audio tones) differently from noise components so at this point the two are separated and, although they go through similar processing, different coefficients and constants are used.
- Calculate the noise masking across each sub-band.
  The next stage is to apply noise masking across the neighbouring sub-bands.
- Calculate the masking threshold for each sub-band.
  This stage calculates the masking threshold for each sub-band and determines the value at which noise is masked; it is used to complete the next stage.
- Calculate the required signal to noise ratio.
  At this point the signal to noise ratio is calculated from the masking threshold. This determines the number of bits that will be used to encode the sample. The amount of compression is calculated. This information is then sent to the bit allocator.

- Send the data to the bit allocator.
  This is where the encoding is actually carried out, based on the samples and the information processed by the psycho-acoustic analysis. The signal to noise values determine the number of bits to be used to encode each sample. The final step is to Huffman encode the bitstream.

Level II

Level II also uses model 1 but is aimed at bit rates of about 128 kbits per second and thus achieves a slightly higher compression ratio. It has a larger sample size than Level I (1,152 against 384). The sample window is also increased to 1,024 samples — but this is not quite big enough to encompass the Level II sample size. To cope with this, two analyses are performed: the first takes the first 576 samples and centres them into the 1,024 sample window. The second analysis repeats the process using the second 576 samples. The results are then combined using the higher of the signal to noise ratios for the final output. This effectively selects the best noise masking values.

The processing for Level II using model 1 comprises the following stages:

- Convert the audio into the frequency domain.
  This is where the audio is converted from the time domain into the frequency domain using a function such as a fast Fourier transform.
- Extract the critical band widths from the spectral information.
  To simplify the calculations, the spectral information is divided into 32 sub-bands, each with a width of 625 Hz. This operation is often performed with the preceding stage using a polyphase filter.
- Extract the tonal (audio) information from the noise.
  Model 1 treats tonal information (i.e. the audio tones) differently from noise components so at this point the two are separated out and although go through similar processing, different coefficients and constants are used.
- Calculate the noise masking across each sub-band.
  The next stage is to apply the noise masking across the neighbouring sub-bands.
- Calculate the masking threshold for each sub-band.
  This stage calculates the masking threshold for each sub-band. This determines the value at which noise is masked and is used to complete the next stage.
- Calculate the required signal to noise ratio.
  At this point the signal to noise ratio is calculated from the masking threshold and this determines the number of bits that will be used to encode the sample. This is where the amount of compression is calculated. This information is then sent to the bit allocator.
- Send the data to the bit allocator.

This is where the encoding is actually carried out, based on the samples and the information processed by the psycho-acoustic analysis. The signal to noise values determine the number of bits to be used to encode each sample. The final step is to Huffman encode the bitstream.

Level III

Level III is the most complex of the three algorithms and is targeted with bit-rates of about 64 kbits per second per channel. It uses the more sophisticated psycho-acoustic model 2 which has several differences. The sample size is the same as Level II and the same split processing technique is used. The processing for Level III using model 2 comprises the following stages:

- Convert the audio into the frequency domain.
  This is where the audio is converted from the time domain into the frequency domain using a function such as a fast Fourier transform. This is the same as Levels I and II.
- Extract the critical band widths from the spectral information.
  To simplify the calculations, the spectral information is divided into 32 sub-bands, each with a width of 625 Hz. This operation is often performed with the preceding stage using a polyphase filter. This is the same as Levels I and II.
- Extract the tonal (audio) information from the noise.
  Model 2 does not separate out the noise for the tonal components (audio) as performed by Levels II and III, but instead calculates a tonality index which indicates whether the component is more likely to be noise or audio. This gives a more accurate separation at the expense of more computation.
- Calculate the noise masking across each sub-band.
  The next stage is to apply the noise masking across the neighbouring sub-bands. With Level III, this is done through the use of a spreading function. With levels II and III, the calculations are based on a set of empirical values. Again, this adds to the computation needed to implement this encoder.
- Calculate the masking threshold for each sub-band.
  This stage calculates the masking threshold for each sub-band. This determines the value at which noise be masked and is used to complete the next stage. Again, model 2 uses a different technique which gives better accuracy at the higher frequency sub-bands.
- Calculate the required signal to noise ratio.
  At this point the signal to noise ratio is calculated from the masking threshold and this determines the number of bits that will be used to encode the sample. This is where the amount of compression is calculated. This information is then sent to the bit allocator.
- Send the data to the bit allocator.

This is where the encoding is actually carried out based on the samples and the information processed by the psycho-acoustic analysis. The signal to noise values determine the number of bits to be used to encode each sample. The final step is to Huffman encode the bitstream.

MPEG-1 systems

So far, this chapter has described how the video and audio data is compressed into bitstreams. The final part of the specification defines how these bitstreams are multiplexed together to create the final MPEG1 bitstream. This is achieved through the systems coding part of the specification. The coding layer specifies a multiplex data format that allows multiplexing of multiple simultaneous audio and video streams as well as privately defined data streams.

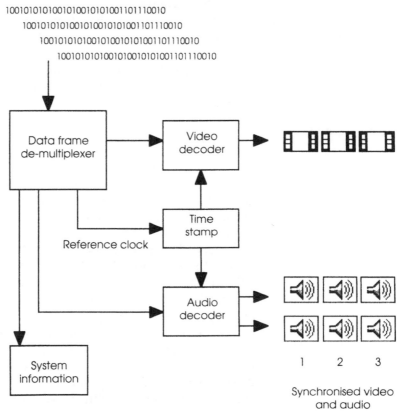

Decoding an MPEG1 bitstream

This systems coding includes necessary and sufficient information in the bitstream to provide:

•   The synchronisation of decoded audio and video frames.

The specification provides a mechanism for the time stamping of frames so that a decoder can synchronise the decoding and playback of audio with the correct video sequence to achieve lip synchronisation. This is done by time stamping the frames using a 90 kHz reference clock. It provides the decoder with great flexibility and even allows the support of variable data rates, where frames can be dropped when they cannot be processed in time. Time stamping provides the reference information to allow the frames to be dropped without losing synchronisation.

- The management of data buffers to prevent overflow and underflow errors.

   With frames not necessarily coming in time sequence, the management of buffers to temporarily hold data is critical in providing a good bullet proof decoder. The systems layer provides support to the decoder for this.

- Random access to frames within the stream and absolute time identification.

   This is important when decoding in that the timing reference can be independent of the environment.

The MPEG1 system is included in the MPEG2 specifications. The next section on the MPEG2 standard covers the role of the system layer in more detail.

MPEG2 overview

As soon as the MPEG1 specification was published, the MPEG committee started to focus on three derivatives called MPEG2, MPEG3 and MPEG4. As it turns out, MPEG3 was eventually incorporated into the MPEG2 specification whilst work continued on the MPEG4 standard.

One of the restrictions of the MPEG1 specification was that it did not directly support broadcast television pictures as in the CCIR-601 specification. In particular, it did not support the interlaced mode of operation, although it could support the larger picture size of 720 by 480 pixels at 30 frames per second.

The problem with interlacing affects the motion estimation process dramatically in that components could move from one field to another, and vice versa. As a result, the original MPEG1 implementation did not handle interlacing at all well, despite some attempts to shoehorn it in. Coupled with this was a need to have better picture quality and to support digital broadcasting and high definition television. As a result, the MPEG2 standard becomes almost all things to all men, in terms of what it could support, while still retaining compatibility with the original MPEG1 standard. To support the wide range of options within the standard for video, the concept of profiles and levels was developed. The different profiles define the algorithms that may be used, while the different levels within the profiles define the parameters that may be used. For example, a level defines sample rates, the frame size and so on and is analogous to the MPEG1 constrained parameter set.

MPEG2 video compression

The MPEG2 specification defines various configurations to provide a set of known configurations for different applications. This is similar to the constrained parameter set within MPEG1 which specified a subset of parameters for implementors to focus on. These configurations are known as profiles and levels.

Profiles and levels

Four profiles are defined and four levels — but not all levels are valid for all profiles. The four profiles are:

Simple
This is the same as the main profile except that B frames are not supported — its antagonists achieved a victory here! It is aimed at supporting software-based applications and decoders and low end cable television.

Main
This is the profile that the bulk of MPEG2 applications and users support. It is envisaged that >90% of MPEG2 applications will use this profile, although this is likely to change when high definition television starts to appear.

Main+
This is the same as the Main profile, except that the encoding is enhanced with the addition of spatial and SNR scalability. These terms are explained later in this chapter.

Next
This is the same as the Main+ profile except that the YCrCb format for the macroblocks is expanded to 4:2:2.

The four levels that the MPEG2 standard specifies are:

Low
This is similar to the MPEG1 constrained parameter set and supports a maximum frame size of 352 by 240 at 30 frames per second (NTSC) or 352 by 288 at 25 frames per second (PAL). This equates to a 3.05 Mpixels per second and a bit rate of up to 4 Mbits per second. It is aimed at the consumer market and offers similar quality to that achieved by a domestic video recorder.

Main
This supports a maximum frame size of 720 by 480 at 30 frames per second as defined in the CCIR-601 specification. This equates to a 10.4

Mpixels per second and a bit rate of up to 15 Mbits per second. It is aimed at the higher quality consumer market.

High 1440   This supports a maximum frame size of 1,440 by 1,152 at 30 frames per second and is four times the frame size within the CCIR-601 specification and used by the Main level. This equates to a 47 Mpixels per second and a bit rate of up to 60 Mbits per second. It is aimed at the high definition (HDTV) television consumer market.

High   This supports a maximum frame size of 1,920 by 1,080 at 30 frames per second and is four times the frame size within the CCIR-601 specification and used by the Main level. This equates to a 62.7 Mpixels per second and a bit rate of up to 80 Mbits per second. It is aimed at the high definition (HDTV) television production (SMPTE 240M specification) market.

	Profiles			
Level	**Simple**	**Main**	**Main+**	**Next**
Low	Illegal	Supported	Main plus SNR scalability	Illegal
Main	Supported	Supported	Main plus SNR scalability	4:2:2 YCrCb
High1440	Illegal	Supported	With spatial scalability	4:2:2 YCrCb
High	Illegal	Supported	Illegal	4:2:2 YCrCb

Profiles and supported levels

MPEG2 enhancements

The MPEG2 standards are based on the MPEG1 standard and backwardly compatible. The basic processes of colour conversion, chrominance sub-sampling, DCT processing and motion estimation on macroblocks are followed but with several enhancements included in the specifications of some profiles and levels.

Many of these enhancements have been incorporated to provide support for new multimedia technologies, such as high definition television and the digital broadcasting of multiple television channels using MPEG2 compression techniques. The main enhancements are:

- All MPEG-2 motion vectors use half pixel accuracy. This is done by taking two adjacent pixels and using the average in the comparison. This increases the computing power needed.
- Alternate scanning pattern which improves coding performance over the original zig-zag scan used in H.261 video conferencing standard, JPEG, and MPEG1.
- Alternative prediction and macroblock modes supported to improve picture quality.

- Chrominance sampling improved as an option within the Next profile by adding support for 4:2:2 and 4:4:4 macroblocks.
- I frame encoding improved — especially its resilience to bit errors — through the use of special concealment vectors.
- Interlaced video using two fields supported.
- Many more frame sizes and aspect ratios supported. A frame size can be as large as 16,383 by 16,383 pixels and different aspect ratios supported providing the vertical and horizontal dimensions are a multiple of 16 pixels. With interlaced video, this restriction is raised to be a multiple of 32 pixels.
- Pan and scanning support to instruct how a decoder should display an image within a window where the aspect ratios are different e.g. a 4:3 image in a wider 16:9 window.
- Support added to allow the source of the compressed video to be identified. This is useful when post-processing the pictures to improve the quality. Information describing the video type (NTSC, PAL, SECAM etc.) and colour primaries (609, 170M, 240M, D65, etc.) can be included. Other characteristics also included which describe aspects of composite video such as v-axis, field sequence, sub-carrier, phase, burst amplitude, etc.
- The macroblock quantisation used in the DCT processing changed to be non-linear and thus offers a greater dynamic range (0.5 through to 56 as compared to 1 to 32 for MPEG1).

MPEG2 scalability modes

The MPEG2 scalability modes are supported only in the Main+ and Next profiles. They are provided to support the prioritisation of video data and hence some level of graceful degradation. The technique is to break the MPEG2 video bitstream into layers and prioritise these layers for transmission. This gives several potential advantages: the highest priority can be transmitted with better error correction and thus ensure its correct delivery whilst the lowest priority data need not be. The receiver decodes the high priority data and is thus to decode the picture but is not totally dependent on the lowest priority data. If it did not arrive, the picture would still be present, albeit with a reduced quality. By separating the video data in this way, there is an increased chance of at least receiving something, as opposed to nothing at all. There is a cost involved with increased bits for the error encoding — but this is not as high as having to impose this on all the video data that is being sent.

An alternative use for scalability is to support the ability to simultaneously broadcast (simulcast) a single bitstream that has the video data encoded so that images of differing resolutions can be decoded. One way to do this is to send the picture using the lowest common resolution and then scale the picture. This has been done in the first diagram overleaf. As a result, the scaling introduces jagged edges to the image. Scaling with additional information prevents this, as shown in the next diagram.

Scaled picture — no additional information

Scaled picture — with additional information

Spatial scalability

Spatial scalability is used to provide support for simulcast where a single bitstream contains encoding for several different resolution picture sizes. The base level is encoded and given the highest priority and higher resolution images (or, to be more exact, the information needed to create them) is sent as additional information within the stream at a lower rate.

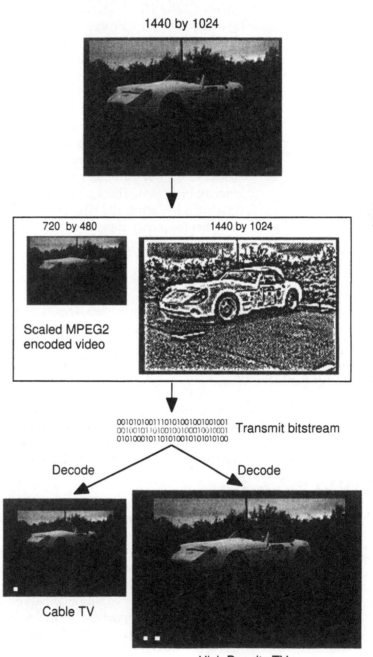

Simulcast using MPEG2 compression

In this way, different decoders can all use and decode the same stream, albeit with different resolution results. The diagram shows how this is done: a large image is encoded into two bitstreams with a low resolution version of the image and the additional information that is needed to construct the full resolution image. The two sets of data are multiplexed and sent out as a normal bitstream. The diagram shows one stream as black 0s and 1s and the other as grey. Both the cable TV and HDTV decoders can understand this scaled video bitstream but they process it in different ways: cable TV simply extracts the low resolution image and discards the additional information whilst HDTV uses all the information to create the full image from the simulcast.

Data partitioning

This technique takes the coefficients from each macro block and splits them into two groups, which are transmitted as separate bitstreams within the multiplexed data. The first group contains the more critical lower frequency coefficients and side informations (such as DC values, motion vectors) and is encoded in the first highest priority bitstream. The second group carries higher frequency AC data in the lower priority bitstream. As the human eye is less sensitive to higher frequency components, the loss of the data from the second group has less effect.

SNR scalability

SNR scalability is a spatial domain method where channels are coded at identical sample rates (but differing picture quality) through differing quantisation step sizes. The higher priority bitstream contains base layer data that can be combined with a lower priority image improvement bitstream to construct a better picture. This can also be used within simulcast technology.

Temporal scalability

This is where video is encoded, usually at a lower frame rate, and the resulting bitstream used to predict the contents of another bitstream. This does not appear to have much use but could form the basis of an implementation of stereoscopic vision where the right channel is predicted from the left channel.

MPEG2 audio

The MPEG2 committee has expanded the MPEG1 audio specifications to form the basis of the MPEG2 specifications. The main difference is in the provision of 5.1 channels for each video channel to support surround sound systems, such as the Dolby Prologic system which augments the more traditional left and right channels of a stereo system with two side channels and a centre channel. All these audio channels support the normal audio range i.e. 20 to 20 kHz. In addition, there is an additional channel that only goes to 100 Hz for use with special effects. This is very similar to the various low frequency audio systems that have been used to enhance the presentation of science fiction and disaster movies. Given suitable power, the special effect channel can convey vibration and movement very realistically. The special effect channel is the .1 or the 5.1 channel nomenclature. The stereo channels left and right are compatible with MPEG1 left and right channels.

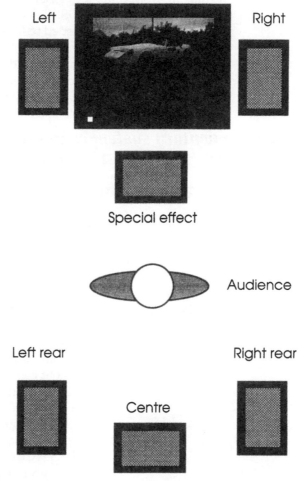

The MPEG2 5.1 channel audio surround sound system

Other enhancements have also been added:

- Half rate sampling rates — 16 kHz, 22.05 kHz and 24 kHz have been included, as well as MPEG1 full rates. These offer improved quality at low bit rates (below 64 kbits per second) at the expense of frequency.
- Additional non-backward compatible encoding techniques, known as NBC encoders, are supported. The Dolby AC-3 is one such NBC encoder that is gaining a lot of support. These have been added to address the lack of a clear encoding method that satisfies all requirements. The methodology behind the NBC is to allow alternatives to be supported, providing they can meet the current requirements as defined by the committee.

MPEG-2 systems

The MPEG-2 systems layer defines how the various bitstreams are multiplexed together and delivered to the decoder. With the standard designed to support many different delivery mechanisms ranging from fibre, terrestrial broadcasting, cable, CD-ROM, other digital storage media (DSM) and satellite, it was clear that two data stream formats were needed to fully address the requirements. As a result, the system layer supports two methods of multiplexing data together: the transport stream and the program stream.

Both streams are constructed using elements called packetised elementary system (PES) which defines the data type and effectively the destination for the data, as will be explained later.

Packetised elementary system components

The MPEG2 standards specify the following PES types:

Video	This is the normal video bitstream which contains the encoded video information for a specific program.
Audio	This is the normal audio bitstream which contains the encoded audio information for a specific program.
Private_1	This is the first private data channel.
Private_2	This is the second private data channel.
DSM	These commands provide control of a local digital storage medium such as a CD-ROM or hard disk.
EMM	Entitlement management message.
ECM	Entitlement control message. This, coupled with the EMM, provides the communication channels for the controlled access to programs either through smart cards or through the use of PIN numbers. For example, a subscriber can ring a supplier, give a credit card number to access a particular program or service, and the access code or encryption key can be sent to the decoder via one of these messages.
Teletext	This provides teletext information for a program. This is normally encoded with the analogue signal in analogue systems. With MPEG1, this is not possible and the teletext information is coded as an external data channel.
Subtitles	This provides subtitling information for a specific program.
Padding	Needed to ensure that the transport packet is always 188 bytes.
Others	For future expansion.

Program service information

The MPEG-2 standards specify the following PSI tables to store the PES program specific information. For more information about their use within an application, see Chapter 13 on digital video broadcasting.

PAT   Program association table
PMT   Program map table
CAT   Conditional access table

The transport stream

This stream is designed for high bandwidth and high error rate transmission, as found with satellite, terrestrial broadcast and cable. It multiplexes many programs into a single stream and thus allows different programs to be chosen from the bitstream.

Transport stream based decoder

It uses PES elements to construct the payload but, unlike the program stream, it uses a fixed size packet of 188 bytes to deliver the data. (Note that 188 byte packet size is also compatible with the ATM packet size.) MPEG transport stream support is seen as a potentially large user for ATM bandwidth. The packets also contain program specific information (PSI) that is used to reconstruct and control the various programs within the stream. Each program — video and associated audio and data channels — has its own clock reference.

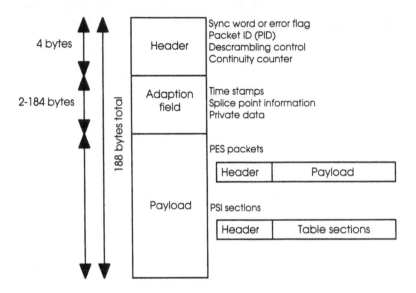

MPEG2 transport stream

The program stream

The program stream is similar to the MPEG1 system but includes some enhancements and extensions to support the 5.1 channel surround sound audio, for example. It was designed for very reliable transmission such as found with DSM and similar systems. It supports a single program, e.g. a video channel with associated audio and data channels, and is targeted at systems where the choice is made by selecting the source rather than contents. For example, to change the movie, the CD-ROM is changed that contains it.

The channels use a single common clock reference for time stamping support. The data format consists of a variable length packet with a header and a payload constructed from as many PES elements as needed.

MPEG decoder

The decoders for both MPEG1 and MPEG2 systems are similar at a block diagram level. Whilst there are obvious differences — the MPEG2 bit stream is a superset of that used within the MPEG1 specifications — the basic process remains the same. The bitstream is parsed and separated out into its different components within the multiplex. The video information is sent to a video decoder and the audio information dispatched to the audio decoder. These two units both decode the streams and are synchronised using the timestamp mechanism to ensure that the audio appears at the same time as the matching video sequence.

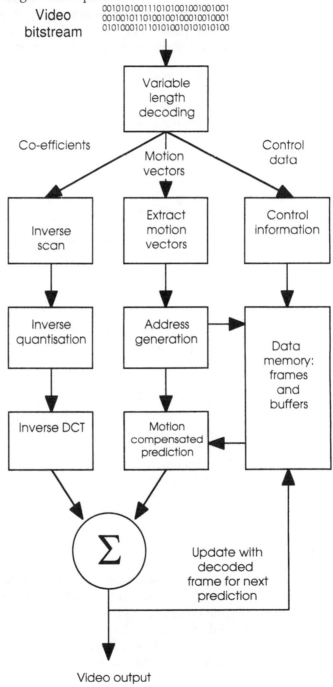

MPEG1/2 video decoder block diagram

The video decoder is a straightforward inverse of the encoding process, as shown in the diagram. It should be remembered that the encode / decode process is asymmetric and that the bulk of the processing power is needed in the encode process; decode requires less resources. One characteristic that this creates is that the quality of video and audio can be affected by the quality of the encode process. If the encode was not carried out appropriately and shortcuts taken, the resultant image from the decoder may not be good.

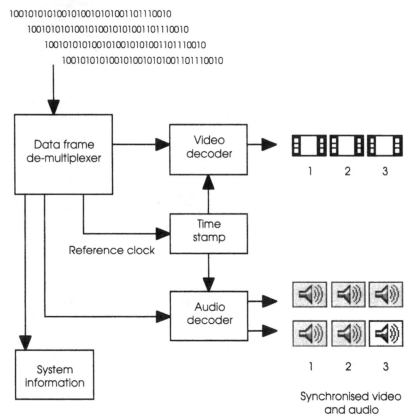

MPEG bitstream

100101010100101001010100110110010
  100101010100101001010100110110010
    100101010100101001010100110110010
      100101010100101001010100110110010

Basic MPEG decoder structure

Indeo

The Indeo compression technology from Intel is a proprietary set of video and audio compression algorithms which is often used in the Windows environment. It is not compatible with the MPEG1 or MPEG2 standards.

Indeo uses the chrominance sub-sampling technique in the same basic way as MPEG but it takes the data decimation a stage further. With the MPEG algorithm the sub-sampling results in a reduction to 25% of the original content. With Indeo, this is further reduced to 6.25%.

Video compression algorithms usually compress data by compressing a frame, extracting the pixel differences between this frame and the next one and then compressing these data using techniques such as run length encoding (RLE). Again, this greatly reduces the data and processing power needed.

Indeo does not use DCT technology but instead uses a technique called vector quantisation. When identifying pixel differences, vector quantisation effectively compares and identifies similar strings of pixels and then assigns an approximate identical value to each one. This reduces the number of differences that are sent and allows further compression at the expense of some data loss.

Video for Windows

Video for Windows (VFW) was Microsoft's response to Apple's introduction of its QuickTime technology. Strictly speaking, it does not describe a compression technology but the framework for supporting different compression algorithms such as Intel's Indeo, Radius' Cinepak and even the Window's version of QuickTime under the Windows environment. It provides a mechanism for the identification of compression algorithms from the format/header of the file which contains the compressed video data and the installation of the corresponding decompression software to playback the video. Decompression can be performed by software alone, in which case the PC performance determines the frame rate and the size of supported picture.

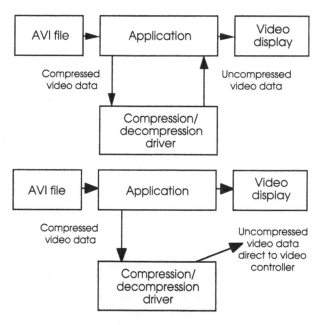

Using a video compression driver

VFW uses a software service called Installable Compression Manager (ICM), which is used by an application or driver to access the compression or decompression routines. A message is sent to ICM describing the required function which then passes this onto the software or hardware that actually does the work. Access through the ICM is assumed to be non-real-time, and the source and destination likely to a file. It also implies that the application is responsible for synchronising audio and video playback if there is insufficient processing power available to keep up at the full frame rate.

The previous diagram shows how the system routes data to and from the compression driver. Although it states that an application communicates directly, it is actually via the MCI interface. Therefore the application is more likely to be a driver that interprets the MCI commands than a true application. The next diagram gives a different view of this mechanism using interfaces and modules.

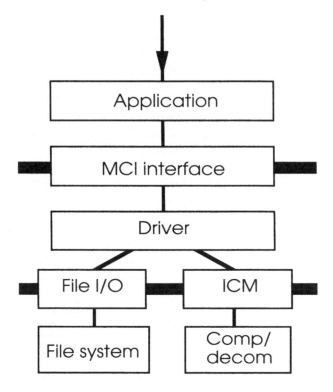

Compression operation

An application may call the compression or decompression routines indirectly by sending an MCI command to playback video to a driver, such as MCIDRV. This in turn would use the decompression module to take the AVI file and decompress it.

The decompressed and compressed video data is supplied via buffers allocated by the application or driver and it is up to this module – not the decompression or compression modules – to route the data to the video display.

The diagrams show combined compression and decompression modules because they are usually combined. This is not mandatory and separate interfaces and libraries are allowed.

The standard modules supplied by Microsoft to support the MCIAVI driver rely on the driver above ICM to supply and remove the frame buffers. This places the responsibility and loading onto the host processor, rather than move it off board. There are options which allow the decompression modules to directly access the frame buffers and draw directly rather than pass a completed frame back to the application who would then draw the picture.

Video for Windows is described in more detail in Chapter 14 on the multimedia PC.

QuickTime

Apple's QuickTime was the first video compression technology to be released as part of the standard operating system for personal computers. It comprises several components:

A QuickTime movie (normal size)

A QuickTime movie (double size)

- Movie Toolbox.
  This provides a set of facilities to allow applications to play-back and create movies without having to write all the code from scratch.
- Image Compression Manager.
  This allows different compression algorithms to be installed and accessed by applications without having detailed knowledge of exactly how they work. It provides a consistent interface for applications and thus shields them from any changes that may be needed when moving from one compression scheme to another. It also resolves issues when playing back movies that were created using one colour depth on a system that supports another e.g. a 24 bit movie will play on a QuickTime system that supports 1, 8, 16, or 24-bits at the same rate as the original.
- Component Manager.
  This part of QuickTime manages the external resources within the Apple MAC and allows hardware such as video capture boards, audio inputs, and so on, to declare their resources and requirements to the Component Manager so it allocates resources as needed. In this way, applications can support new hardware without modification.

Photo compressor

Apple's photo compressor is based around the JPEG standards and provides JPEG compression and decompression support as part of the standard operating system release. One advantage of this is that many applications can decode JPEG pictures that conform to the Apple PICT JPEG format without having to implement the decoder as part of the application. The Simpletext editor, for example, is capable of such decoding through the use of the built-in QuickTime photo compressor.

Animation compressor

Apple's animation compressor uses a compression algorithm based on run-length encoding to compress computer-generated animation. The implementation supports both loss and lossless compression, depending on the target performance required and the processing power needed to decode the compressed animation.

Video compressor

Apple's video compressor uses a proprietary image compression method developed by Apple. The technology involves the use of both spatial and temporal redundancy, i.e. chrominance sub-sampling and motion estimation, but the actual details are closely guarded. The algorithm has been ported to both the MacOS and the Microsoft Windows operating systems. It provides compression ratios typically in a range between 5:1 and 25:1. One advantage the algorithm provides is a relatively low level of performance needed to decode the video and play it back. A typical QuickTime movie which was compressed using this algorithm can achieve play back rates of 15 frames per second with a picture size of 160x120 pixels on a 20 MHz MC68020 based Macintosh. Larger picture sizes and frame rates are supported by faster processors such as the PowerPC based MACs.

Graphics compressor

The Graphics Compressor also employs an image compression method developed by Apple. It provides lossless compression of 8-bit images and is ideal for compressing both still images, such as those created in painting applications, and 8-bit movies. The Graphics Compressor differs from the Animation Compressor (which can also compress 8-bit data), in that the Graphics Compressor gains compression at the expense of decompression speed. A movie compressed with the Graphics Compressor will usually be half the size of an Animation Compressor (RLE) movie at approximately half the maximum playback rate.

Compact Video Compressor

The Apple Compact Video Compressor is an advanced version of the Video Compressor. It increases the image quality, playback size, frame rate and compression of digitised video compared to the original Video Compressor. It can play back a movie on an MC68020 based Macintosh running at 20 MHz at 240x180 pixels at 15 frames per second with the source material accessed from CD-ROM. With more powerful Macintoshes, frame sizes and frame rates are further increased to 320x240 pixels and 24 frames per second.

Wavelet compression

Wavelets first appeared in the 1980s in geophysical work but the technology quickly spread to other areas of applied mathematics. The fundamental premise behind is that any signal can be represented by combining many simpler wave forms — called basic functions — using weighting factors — known as co-efficients. The Fourier transform which will decompose a signal into a set of sine waves of specific frequency is a well known example of this approach. With wavelets, the basic functions are not retricted to sine waves as is the case with the Fourier case and can be quite different in form and thus their characteristics.

Wavelet compression is in some aspects very similar to the DCT compression techniques that are used with MPEG and H.320 video conferencing. It transforms the data to make it easier to decide which data can be lost for the minimum impact on video quality and then uses coding techniques to compress the data. The big difference though is that the DCT algorithm is not used and is replaced with a wavelet transform. The wavelet itself can vary and several candidates have been put forward.

Using wavelets gives better image quality at low data rates and higher compression when compared to the DCT-based compression. The wavelet can be chosen or adapted for the required characteristics. For the viewer, this improvement takes the form of more pleasing artifacts: with DCT systems these artifacts appear as blocks that seem to stand out from the rest of the picture. With a wavelet based system, the block based artifacts are not created and all that is seen is a progressive loss of detail as the compression ratio is increased. This advantage has prompted a lot of interest in wavelet technology for video compression. A commonly used one is the bi-orthogonal wavelet transform. This uses both high and low pass filters and downsampling to transform the data. The process is iterated several times to get the required detail.

The bi-orthogonal wavelet transform uses a set of filter banks to transform the original signal into a set of filtered images where each image contains less and less high frequency data. This is done by using a filter tree that filters the original image and subsamples it to create four images, with each image being half the size of the original. This is typically done by using a combination of low and high pass filters. One of these images will predominantly contain the low frequency information which the human eye is better at resolving compared with the high frequency information. This image is then processed again to further resolve the information and so on. This is performed on the Y, Cr and Cb video components to create a complete tree. This information is often depicted as a montage of all the resulting images. It is important to realise that the output from each filter is then used to feed another iteration and the end result is a set of filtered and scaled down images. The number of iterations can be restricted or expanded if necessary, depending on the requirements of the compression and to some degree the content of the picture being compressed.

The arrangement for a filter tree

The filter tree output for an image (simulated)

Y transformed data

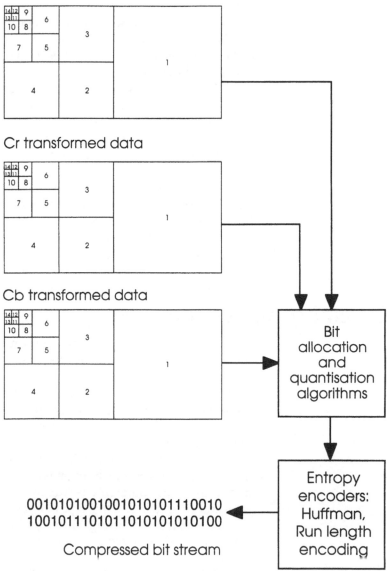

Cr transformed data

Cb transformed data

Bit allocation and quantisation algorithms

Entropy encoders: Huffman, Run length encoding

00101010010010101011100101
10010111010110101010100

Compressed bit stream

Wavelet based video encoder

Once the filtering has been completed, the compression i.e. the run length encoding techniques can be applied. Up to this point no compression has been performed because all that has been done is to transform the data into a format that allows the compression techniques to be better applied. i.e. identification of the different components in the picture. The next part of the process is to quantise the data and apportion data bits to these various components.

The smaller images are more important because they contain more information about the lighter parts of the picture. The larger images contain less and therefore data can be lost here without it being so noticeable. The end result is that dark areas of data are compressed more than the lighter and more colourful areas to exploit the fact that the human eye is less sensitive to dark colours compared to light ones.

Once the quantisation and bit allocation have been completed, the data can be further compressed using Huffman or run length coding to create the final compressed image. Image decoding is the reverse of the coding process.

Wavelet coding is becoming more popular especially as more processing power is now available on desktops to implement software based codecs and due to the introduction of hardware codecs. It is likely that it may be used in areas where standardisation is not an issue or where new standards are being defined. Unfortunately as a technology it is too late to change the MPEG standards and because of the investment in compressing the source material (films and videos) and the advent of low cost decoders, MPEG and the use of DCT-based compression techniques are likely to continue into the next millennium. It is noticeable that there are wavelet based plug-ins now available for Internet browsers and thus it is likely that the PC and the Internet will benefit from the introduction of wavelet based technology.

Fractal compression

Mention the word fractal and for many people it conjures up the image of either a computer generated leaf or cloud or an abstract colourful image. These images have some interesting properties. Each time the picture is magnified, the more detail is seen. If this is performed with a bit map image, all that happens is that less detail is seen and this will continue until only a single pixel is seen as a single block of colour.

The word fractal was coined by Benoit Mandelbrot to describe a fractured structure which contains many similar looking structures and forms. This commonly occurs in nature in snowflake crystals, trees and river deltas. These type of images can be created by taking a simple structure and using this as a building block. A Y-shaped twig can be expanded or reduced, rotated and added to the picture to create the required image. This means that a basic shape can be reused to create the new image using a set of coefficients and equations to control the basic image manipulation. These take the basic image, manipulate it and then lay it onto the image to gradually build up the picture.

The manipulations that can be performed are rotation, skewing, scaling and translation. These are used to create new derivatives of the original image which can be used to build the final image. This gradual build up is referred to as collaging and the image manipulations are called affine transformations.

Original   Skewed   Rotated   Scaled

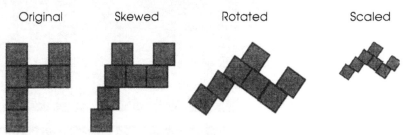

Fractal transformations

Many natural images can be created through collaging using affine transformations. Many algorithms have been developed to find efficient methods of doing this. The use of fractals as a compression technology is based on the idea that if fractals can be used to create images that are extremely lifelike and natural, then these images can be compressed by discovering the fractal transformations that can be used to replicate these images. In other words if a few equations can define and create a fern leaf, then a picture of a fern leaf could be compressed to a few equations.

This idea is further enhanced by the fact that many real life images have a tremendous amount of redundancy within them. This idea started to become a practical realisation in the late 1980s when Michael Barnsley and Alan Sloan described how a collection of affine transformations could create a replica leaf by applying an algorithm called the Chaos Game. They went on to found Iterated Systems who have developed compression technology based on fractals that could automatically compress images. The most commonly used application is the Microsoft Encarta multimedia encyclopedia program that uses Iterated Systems' fractal technology to compress the pictures that are stored on the CD-ROM. One of the disadvantages of fractal technology is that it is assymetric. The encoding process is far more processor intensive compared to the decode process, although this is becoming less of a problem as CPU power continues to increase.

Fractal technology is also used to create artificial 'natural' landscapes and drawings. By using standard fractal transformation to create natural looking textures and combining them with 3D modelling techniques, it is relatively easy to create realistic looking scenes without having to draw the textures pixel by pixel. The drawings can be stored in their natural fractal based formats and expanded to any required resolution. An example of this technology is shown overleaf.

An artificial landscape created using a fractal-based drawing program. The top picture is the result of just two iterations. The bottom after 50.

The fractal compression process

The compression process comprises of several stages. The first stage is to split the image into a set of adjacent and non-overlapping areas called domain regions. This is similar to splitting an image into rectangular blocks except that these regions can be any shape or size. One simplification is to limit the domain shape to a regular shape such as a triangle or rectangle. The next stage is to create a set of range regions which need not cover the whole image and can overlap. The range regions are then analysed to identify how they can be made to resemble the domain region. Each range region is a potential candidate and is transformed using an affine transformation and compared. Included in the transformation are adjustments to the contrast and brightness.

Once the matches have been found, the image can now be described as the regions and affine transformations that are needed to create the collage.

The challenge for fractal compression is in using the right — most efficient or suitable would be a better description — regions and domains along with the affine transformations. The number of possibilities are almost endless but this can cause practical problems in terms of the processing and time needed to perform the compression. Commercial systems tend to use boundary conditions such a matching criteria, time and processing constraints to limit the complexity of the compression process. Other analytical techniques can be used to help identify likely candidates for the domain and range regions. A large domain field may be ideal for an area or blue sky or a cloud but unsuitable for a small detailed object where a small domain is a better choice.

Once this has been completed, the chosen domain regions along with the affine transformations are then stored in a file to create the fractally compressed image file. Note that the compression process is different from that used in wavelets and DCT based algortithms, in that the compression has been achieved by identifying a set of fundamental relationships that can be used to derive the final image. Wavelets and DCTs transform the image information into a format that is more efficient when used with traditional data coding techniques such as run length coding.

The fractal decompression process

The decompression process is a little strange at first. It starts by taking an arbitrary image and using this to generate two images A and B of the required size. The image content at this point is immaterial and can be anything at all. The next stage is to assign one image to be the range image and the other to be the domain image. In this example, image A is the range image and image B is the domain one. A is then partitioned into the domains based on the domain information from the fractal file. For each domain image, the affine transformations are then extracted from the file, applied to find the range regions and then mapped to image B to update the domains. Once this has been done for all domains, image B has been partially transformed into the decompressed image. The first iteration of the decompression has been completed.

For the next iteration, the images are swapped over with image B now becoming the range image and image A is now the domain one. The process is repeated again and this time image A is updated. The iterations are continued until there are no differences between images A and B and then the image has been decompressed. The number of iterations is dependent on the content and the final picture size.

The quality of the final image depends on the accuracy of the affine transformations and thus the limiting criteria used to limit the compression analysis. Fractals have promised a lot initially but it is only now that they are delivering. They can achieve good compression. They, like wavelets, have more pleasing artifacts that do not show the blockiness that DCT based compression algorithms can exhibit, especially at low bit rate encoding. As smarter and more efficient algorithms and faster processors become available, the main problem of the compression time will disappear and fractals may start to become a serious challenge to DCT based compression.

MPEG 4

The MPEG 4 specification was initially designed to support very low bit rates and is designed to send video down telephone lines. An ideal application for this would be video phones. This almost competes with the H.320 video-conferencing standards, which already support audio plus video communication down telephone lines and has derivative work being carried out on even lower bit rate encoding. The initial work — the committee has only just recently completed the work and the first complete specification is not expected until 1999 — supports frame sizes up to 176 x 144 at 10 frames per second and uses a bandwidth between 4.8 and 64 kbits per second. This has been expanded to encompass digital television and other types of digital audio-visual communication at even higher bit rates.

The actual approach taken with MPEG 4 was to move away from the frame approach to video compression where the whole frame is processed as a whole but to one where the individual components that make up the video such as people, objects and backgrounds are identified and treated separately. This approach also includes the ability to include synthesised objects as well as synthesising movement.

The approach is applied to a multimedia stream which encompasses both video and audio, presentation material and so on. The material is processed to identify the individual components which are referred to as audio visual objects or AVOs. These are arranged in a hierarchy. Simple examples of this would be background, foreground objects, people and speech. The AVOs are related to each other and these relationships are captured within the hierarchy. In addition to these objects, others are also supports such as text and graphics, animated bodies where the movement is synthesised and talking heads where both the facial expressions and speech are synthesised from text.

The objects are treated as individual entities: people are defined on their own with no background at all. The same is done with objects. This allows the basic information to describe a scene to be sent once. As people and objects move in the scene and thus reveal and hide parts of other objects and background, this information does not need to be sent as the data that describes these objects is already present. It could be argued that this is no different from other compression techniques where only the data that changes is sent. This is a valid point except that with MPEG 4, it is only the moving object that is considered and that other visual changes — hiding and re-appearance — are not considered. This also leads onto several other interesting aspects.

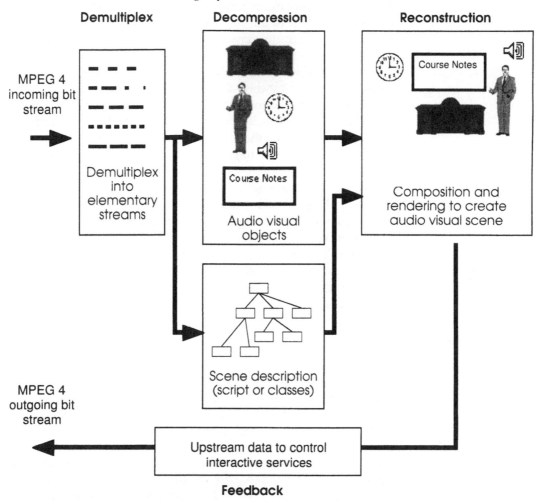

Basic MPEG 4 bit stream processing

The basic processing is shown in the diagram: a multiplexed bit stream is received and demultiplexed into MPEG 4 elementary streams. These streams can contain information to define either audio, visual or synthesised objects. These objects are then decompressed to create the objects that will be used to create the final scene. Included in the elementary streams is a scene description which defines where the objects are in the final scene. The combination of the objects and scene definition is used to create the final scene. There is also an upstream path which allows the viewer to feedback information to the sender to provide interactive control. This feedback information is optional or even may use a different data communication path or method from that used to deliver the original bitstream. It could be a telephone connection with the incoming MPEG 4 stream supplied by satellite.

If the picture is coded by decomposing the picture into its constituent objects, then adding synthesised objects is also very easy. In this way, other information can be included and appear to be part of the overall picture. For example, a computer based presenation could be inserted. In the diagram, the text "Course Notes" could be replaced by a computer generated presentation. Pointers and text can be added to annotate the current picture.

This can be taken even further by using animation and synthesis techniques to synthesise objects or animate them based on a simple initial description. A single still image of a human head and shoulder could be sent and the facial movement generated during speech could be synthesised. This is the idea behind the talking head approach. At the start of a conversation, each party sends their picture. This is received and used to generate along with a selected background the video image. At this point, video is no longer sent but speech is. This is then analysed at each end and used to generate the facial expressions with which to animate the still picture that was originally sent at the start of the transmission.

This process can be expanded further by sending text instead of coded speech and using a speech synthesiser to generate the speech. This is also used to generate the facial movements and the end result is a synthesised talking and moving person despite the fact that no real video or speech is transmitted.

The compression path

MPEG4 should be viewed as a collection of compression techniques that can be applied depending on the bit rate available for the compressed stream and the content of the audio and visual components. The compression path consists of three main parts. The VLBV (very low bit rate video) core provides support for applications that operate at data rates below 64 kbits/s with low frame rates and small frame sizes. The same basic functionality is also provided in a HBV (high bitrate video) core to support conventional MPEG 1/2 applications that uses far higher data and frame rates with larger frame resolutions. The third part is used to support and process the content based support and performs shape processing before using the more DCT based compression algorithms. The shape processor extends both the VLBV and HBV cores to support AVOs and allow hybrid synthesis where natural and synthetic objects can be combined.

As a result, an MPEG 4 encoder or decoder looks very similar to that of an MPEG 1/2 system at a high level except that it has this shape processing block to extend the VLBV and HBV functionality. It uses the DCT with runlength encoding as the basic compression engine but it adds a further block to handle objects. This shape processing block uses several new techniques to handle visual objects, especially those that are not block based and require the use of a mesh to break the object into small areas that are suitable for compression.

Natural audio objects

The MPEG 4 standard supports the normal audio compression techniques used by MPEG 2 but these assume the availability of fairly high bit rates. To allow audio to be supported at lower bit rates, several different audio compression algorithms are needed to encode audio using different bit rates. These techniques are commonly used in other digital communications systems such as cellular telephones, video conferencing and so on. For bitrates of 2 to 4 kbits per second using a 8 kHz sampling rate which is adequate for speech, a parametric coding system is used where the speech is translated into a set of parameters which are sent and then used to reconstruct the speech. This technology can also be used for audio with a higher bitrate of 4 to 16 kbits per second and an optional sampling frequency of 16 kHz as well as the usual 8 kHz. Bear in mind that at these low bit rates and sampling frequencies, the quality is not very good but it is acceptable for communications.

For medium bitrates, speech is encoded using CELP (code excited linear predictive) codecs that use an electronic version of the human vocal tract to recreate the speech by sending input waveforms into the model to excite it and then modifying the sound in a similar way that the movement of the mouth and tongue can change the basic vibrations that the vocal tracts make. These type of codecs are used in cellular telephone systems because they offer reasonable quality at these low bit rates. It should be remembered that 8 kHz sampled speech requires 64 kbits of data per second with an 8 bit sample size. This is at least an order of magnitude higher than the low bit rates that are specified for speech. In this way, the audio can be compressed. MPEG 4 specifies two sampling frequencies at this bit rate to support both narrowband speech and wideband speech. Narrowband speech is the typical quality that is provided on a normal telephone system where the frequency range is approximately from 100 Hz to 3.1 kHz. Wideband speech extends this further to about 8 kHz. The speech quality is much improved with high frequency resonances associated with the letter S for example becoming clearer.

Two further codecs, known as TwinVQ and AAC, that use time to frequency techniques are used for audio for higher bitrates from below 16 kbits per second to 64 kbits per second.

Synthesised audio objects

As an alternative to natural sound, the MPEG 4 standards also support synthesized sounds which in some cases allow speech and other audio to be compressed to allow it to be sent using extremely low bit rates. Speech can be synthesised by sending text which is then input into a text to speech (TTS) module that takes the text and generates speech. In addition to the text itself, additional parameters can be sent to allow the synthesised speech to be tailored to reflect the source in terms of age, gender, language and even the dialect. This allows some interesting possibilities in that these parameters can be changed based on the preferences of the other party or even the content of the material.

The TTS module can also provide facial animation information to allow the synthesised speech to be linked with facial animation to create a complete synthesised talking head.

The second synthesis tool is a special language called SAOL (structured audio orchestra language). It is used to define a collection of audio entities that control and process audio signals. These are referred to as instruments and are analogous to the collection of audio blocks in a synthesiser that can create a particular sound. The *instruments* are (naturally) grouped to create an *orchestra*. The language does not describe the actual method but uses current techniques such as wavetable, FM synthesis and other techniques. It is designed to be expandable to include future techniques as and when they are developed and become available. It will also support audio effects such as reverb and flanging, chorus and so on. Instruments and orchestras do not have to be resident on the receiving terminal but can be downloaded if required.

This does assume that the terminal has the correct support. This may simply be a sound card and a large amount of processing power. In this case, software can be downloaded to create the correct support that is needed. The SAOL does not know or specify exactly how the terminal will create the *instruments* and/or *orchestras*. However, to ensure that all compatible terminals can reproduce the audio without requiring all terminals to have the same level of sophistication, a basic wavetable bank format has been defined to allow simple wavetable-based functions and samples to be downloaded and thus allow reproduction. This does not prevent far more sophisticated facilities to be used if needed but does ensure that all terminals can playback the audio stream with all components in the right time frame. This may mean that on simpler terminals, the level of reproduction quality may not be so good as that on a more sophisticated system.

Synthetic Objects

The standard defines several groups of synthetic objects that can be used to create visual images locally, based on information contained in the MPEG 4 data stream. These can be combined with real live video to create scenes.

Facial animation

Facial animation is used to generate synthetic talking heads without continually transmitting live video as is normally done with a video conference system. The information to define the face is sent as facial definition parameters (FDPs) which are then applied to an initial generic face to create the required face. The facial animation is controlled by information sent as facial animation parameters (FAPs) which are used to control the facial movements such as the mouth movements when speaking.

The initial face is capable of receiving and processing FAPs as soon as it is created. This means that there may well be some morphing appear as the generic face is modified by incoming FDPs.

Rendering the Face object with FDP information to create a synthesised natural Image (simulation)

There are also several levels of local control: FDPS can be generated locally and used to modify the face without requiring FDPs from the remote party and the incoming bit stream. There is a local control that allows the animation to be exaggerated or amplified based on a set of parameters called the amplification factors. This can be used to exaggerate the lip movement in speech animation to help lip reading. Finally, there is a set of local filter factors that allows incoming FAPs to be filtered before being applied to the face object.

Body animation

This is similar in many ways to the facial animation previously described. It defines a basic human body visual object called The Body which is defined using a set of 3D polygon meshes. This generic object can be customised using information contained in body definition parameters (BDPs). This allows the body to be textured and sized so that it can look like a person or a character. The animation is controlled by data supplied as body animation parameters (BAPs). The default posture for the body is standing upright, feet forward with the arms down by the side with palms facing inwards.

2D animated meshes

The video compression schemes that have been described so far break up the picture into regularly sized blocks. Unfortunately, many natural objects such as people do not easily partition into block sized components. As it is more likely that video objects are irregularly shaped, the use of an alternative partitioning could be benficial. MPEG 4 has adopted a 2D mesh using triangular patches. The meshes can be used to define textures for sythetic objects or even used for compression by sending the mesh and then animating it locally.

Generic 3D meshes

Three-dimensional meshes are also supported for more complex representations. The standards define a set of toolbox facilities for the efficient compression, rendering and manipulation of these meshes.

Error recovery

One of the problems with the MPEG 1 and 2 standards is that they assume that the data delivery mechanism is error free. In many cases, this is a correct assumption if the data is coming from a CD-ROM or hard disk. Other sources such as broadcast services use their own error protection wrappers to encapsulate the MPEG information from delivery errors. This reduces the available bandwidth that can be used for the video and audio compression but ensures that the programmes can be viewed without errors or similar interruptions.

A 2D mesh to define a head and shoulders object

MPEG 4 has specified several data recovery techniques to help cope with errors in the delivery system. The aim is not necessarily to recover lost data but to mask it so that its loss is at best unnoticed and at worst, as small as possible. One interesting technique is the use of reversible variable length codes (RVLC) which gives data recovery tools additional opportunities to recover data. These allow data to be recovered from RLE and Huffman encoded sequences that have errors. The process relies on two aspects: the first is that the data stream is resynchronised after the error to detect the end of the code. The tool now knows where the start and the end of the code begins and thus it can work forward to decode and recover data. With a normal code this is all that can be done. With a reversible code, the process can be applied to the end of the code by working backwards. This means that the data recovery can work on either side of the error and thus have a better chance of recovering data.

If data has been lost or corrupted and cannot be recovered , then the next step is to conceal the errors so that they are not visible. With video data, there is considerable amount of data that can be used. One technique is to substitute the block from a preceeding frame in place of the corrupted one. With the frames being updated frequently, this can be quite acceptable. Similarly if a motion vector is corrupted, it can be estimated from previous data and this value used instead. These types of techniques are already used in some implementations to improve the quality of the decoding and minimise any errors.

To further improve this technique, the MPEG 4 standards specify the use of another marker to indicate when the data stream's contents change from motion vectors to blocks. This allows errors in one set of data to be discarded and replaced by information from the other set. For example, corrupted blocks can be replaced with motion vector data and so on.

MPEG4 challenges

The biggest technical challenge facing MPEG4 is how to decompose a video picture into the AVOs. Video pictures are made from a set of pixels and the relationship between pixels is not apparent from the pixels themselves. To derive the objects, object recognition is required. This is something that humans are extremely good at but computers are not. This is extremely difficult and processing intensive, especially where the boundary between the objects is not clear where they have similar colours.

The second problem concerns the fact that the background and object descriptions will be incomplete. The first time an object moves, it will reveal more information about the background and objects behind it. This information will need to be captured, decomposed and then used to enhance the original descriptions.

Other problems can occur when an object that was part of a background becomes an object in its own right. Taking a book off a shelf is a good example of this.

Background objects

Original frame

Foreground object

Decomposing a picture into visual objects

Is this really video and audio compression? This is not an easy question to answer. When the source material is real, then it undoubtedly is but if the picture is completely synthesised so that the only original information is text and possibly a still image, then this becomes a bit blurred.

MPEG 7

Although this standard has been given an MPEG designation, it is not a video compression standard but a language definition that is used to describe audio-visual objects and how they can be decoded.

6   CD-ROMs

CD-ROMs are a spin off from the Compact Disc (CD) technology that has revolutionised the audio world. Instead of storing digital audio data, files are encoded, giving some 600 Mbytes of storage on a single Compact Disc. CD-ROM drives use the same principles as a Compact Disc player – a laser generates a fine beam of light which focuses onto the tracks containing the data. The reflective surface returns the beam except where a pit has been etched into the surface – this allows binary data to be stored and read back.

The original audio based technology was developed by Philips and Sony in the late 1970s and culminated in the issue of the Red Book specification for audio CDs in 1982. The potential application of CD technology as a high-capacity, low-cost medium for read-only data storage resulted in the 1983 Yellow Book CD-ROM specification. Whilst the basic technology remains the same as that for CD audio, CD-ROM data requires greater data integrity: a corrupt bit that is not noticeable during audio playback becomes intolerable with computer data. In addition to the Red Book CIRC (cross-interleaved Reed-Solomon code) standard for audio CDs, the Yellow Book specification addresses the data integrity issue and dedicates more bits for EDCs (error-detection codes) and ECCs (error-correction codes).

These devices are effectively read only and can only transfer data at the same speed as a floppy disk. Some CD-ROM players can double as players for audio Compact Discs. For someone considering buying a CD-ROM and a CD player, this could be the ideal solution.

CD-ROM technology

A standard CD is made from a transparent polycarbonate substrate which has a set of pits embedded into its surface. It is these pits along with the land areas between them that encode the data on the disc. This is then covered with a layer of aluminium or similar alloy to provide a reflective surface. A lacquer is then applied to provide a layer of protection followed by the CD label. The data is read from the disc by sending a beam of light which is reflected back if it lands on a flat area of the track or dispersed (and thus not reflected) if it hits one of the 120 nm deep pits. In this way, digital data is stored using the land areas (known as lands) and pits on the disc.

The data is stored in a single spiral track that has a width of about 600 nm with adjacent tracks about 1.6 μm apart. This is similar to a record, except that the start of the spiral is at the centre and not at the outer edge. With most CDs, this provides an advantage as the outermost five millimetres of the disc are left blank and are used to provide a handling area. With an extended play disc that gives up to 74 minutes playing time, this handling area is used for data and these types of disc are thus more susceptible to damage through mishandling.

The spiral track length with a 120 millimetre wide disc is about 5 kilometres. The track is divided into equal length sectors that can hold 2,048 bytes of data. The disc is rotated at a constant linear speed so that pits and land areas pass through the detector at the same speed, irrespective of where they are on the spiral. As a result, the disc's rotational speed is changed depending where the data is read from. If the detector is reading from the edge, the rotation speed is slower compared to when reading from the start of the track at the centre of the disc.

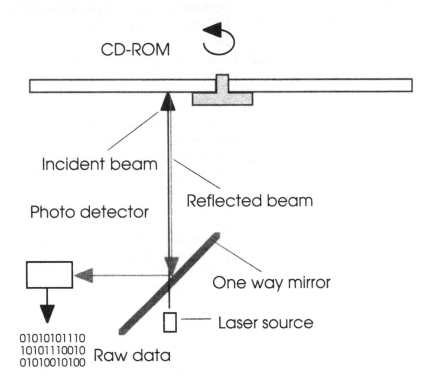

CD optics

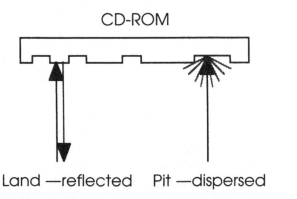

CD light flow

The optical mechanism uses a laser diode to generate laser light that is focused on the disc. The light passes through a one way mirror. If the light hits a land area, the reflective coating reflects the light back down the path it came from to the mirror. It is then reflected again due to the one way characteristic of the mirror and it strikes a photo detector, such as a photo-diode. If the incident light hits a pit, it is dispersed and no reflected light hits the detector. In this way, the diode is turned on and off by the reflected light. The electrical signals it produces match the organisation of pits and land areas on the disc and provide the raw data stream from the disk. It is important to remember that this raw data stream does not directly represent the 0s and 1s that would normally form the digital data.

Raw data organisation

Although the pits and land areas are easily interpreted as directly representing 0s and 1s, this is not the case. The reason is complex but is based around the problems encountered with very large amounts of data that are constructed out of repeated bytes.

The problem concerns knowing when a pit is a pit and when it is a pit that is bigger than expected. If you imagine a data structure of a large number of zeros, the pit area would be very long and to read the number of zeros that are stored, the rotational speed of the disk would need to be either very accurate or accurately measured. This is necessary to generate a clock that can be used to count the number of pits present. The clock is used to synchronise when to read the disc and determine if there is a pit or land area. Whilst this sounds fine on paper, it presents immense technology problems in achieving such accuracy. The clock in such an example would have to be fast enough and accurate enough to measure a distance of 300 nm — not something that can be easily manufactured today for the consumer market.

To provide a more resilient system, the data encoding does not use a straightforward direct relationship between the pits and data bits. Instead, a more complex technique is used that effectively generates its own clock and keeps the maximum pit length (i.e. the number of consecutive pits and lands) to a small minimum. The approach taken uses the following techniques:

- Each transition between a land area and a pit represents a *one* transition. This is raw data and does not directly represent a stored data bit. This has the added benefit of generating a clock edge from the transitions directly from the disc and the drive mechanism. This reduces the complexity of the drive mechanism and allows greater mechanical variations to be tolerated.

- Each continuation of a pit or land represents a zero. Again, this is raw data and does not directly represent a stored data bit.

- The minimum size for a pit or land is 900 nm, which means that consecutive one bits are not encoded or supported.

- The maximum size for a pit or land is 3,300 nm, which means that 10 consecutive one or zero bits are not encoded or supported.

- Two consecutive one bits must be separated by at least two zero bits to meet the minimum pit size.

- Each data byte is encoded using 14 bits on the CD ROM using a technique called EFM (eight to fourteen modulation).
- The data is also run length encoded. To prevent run length encoding clashes between adjacent 14 bit words, three merging bits are used to separate the 14 bit structures.

These techniques are combined to provide the encoding mechanism. The pit-land transitions are used to generate a clock using clock recovery circuits. This is usually a phase locked loop to generate the clock. The transition signal is used to provide the feedback to synchronise the clocks. In addition, the transitions provide the presence of a one bit on the disc. The recovered clock is used to read or decode zeros on the disc by effectively measuring the length of the land or pit area. These areas are artificially kept large by defining a minimum length of 3 pits or lands. At this point, the raw data can be read from the disc.

The EFM takes the 14 bit structures, which represent over 16,000 possible combinations, and allocates 256 codes to represent a stored data byte. Each one of these 14 bit codes meets all the requirements above. As the 14 bit structures are read, the controller decodes it using a lookup table and passes the data byte to the outside world.

Sector organisation

CD-ROMs use a 2,352 byte sector whose organisation varies, depending on the intended use of the disc. A standard CD-ROM supports three modes for the basic structure as shown in the diagram.

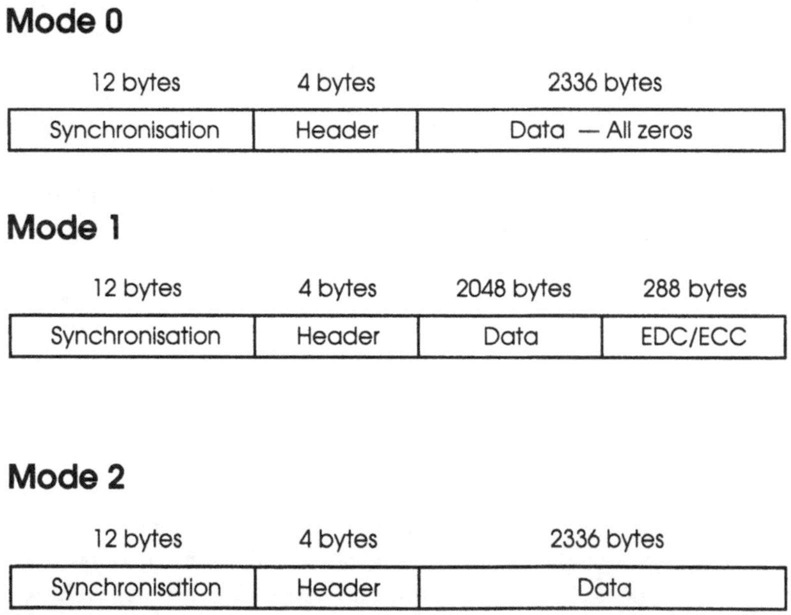

Mode 0

12 bytes	4 bytes	2336 bytes
Synchronisation	Header	Data — All zeros

Mode 1

12 bytes	4 bytes	2048 bytes	288 bytes
Synchronisation	Header	Data	EDC/ECC

Mode 2

12 bytes	4 bytes	2336 bytes
Synchronisation	Header	Data

CD ROM sector structures

Mode 2 allows more data to be stored but does not have any error protection. This is often used for digital audio or video, where odd bit and byte errors are largely not noticeable. Mode 1 is normally used for computer data, where a bit or byte error is critical. Some data storage is allocated to an error correction code (ECC) that can be used for error detection and correction (EDC).

Mode 0 is a third special mode where all the data is set to zero. This is frequently used in uncompressed digital graphics where large blank areas of the picture will simply be filled with zeros. This mode is becoming less important in this particular area because most graphics images are stored using some form of compression, which reduces the blank areas to a small non-zero code.

As a result of the modes with and without error correction, CD-ROMs can have different data capacities. These differences are further increased by the use of the extended play CDs that use the 5 mm outer handling edge as well as the normal data area. By using the whole area and error correction, the disc can hold 742 Mb of data, which equates to audio playing time of 74 minutes. A 60 minute CD holds about 527 to 540 Mb of data using error correction and 580 to 601 Mb depending on whether the megabyte is defined as 1024 bytes or as 1000 bytes. Specmanship is still alive with mass storage devices!

Multi-speed drives

The typical transfer rate for a CD-ROM is about 150 kb per second and, although faster than a floppy disk, this is considerably slower than that of a hard disk. The reason is due to original CD-ROM specifications, which defined the rotational speed and thus the rate at which data could be read from a CD-ROM disc. Whilst the addition of disc caches greatly improved the transfer rate (especially with random data searches by storing whole tracks of data for later access in the cache memory) a new technique appeared which speeded up the rotational speed and thus increased the data rate. These multi-speed drives spin the CD-ROM at 2, 3 or 4 times the typical speed. These double, triple and quad speed drives offer far faster data transfer rates which greatly improves data access and multimedia performance, particularly when playing back movies. They retain compatibility with audio discs by recognising audio tracks and restricting the rotation to the normal speed. Some models even allow direct control of audio discs from the front panel.

Multi-speed drives have become the norm and are considered to be the minimum for multimedia applications. As a result, single speed drives are becoming obsolete. However, for use as reference drives or to allow the installation of software and for playing audio CDs, they can be a very cost effective addition to a system. Single speed drives can often be obtained for not much more than the cost of a floppy disk drive and are frequently bundled with a SoundBlaster or equivalent card, speakers and CD-ROMs.

CD-ROM IBM PC interfaces

There are almost as many CD-ROM interfaces as operating systems available for IBM PCs and this can cause problems unless the

CD-ROM drive is bought for a specific interface or with its own interface card. Unfortunately, many of the interfaces share the same connector and thus cannot be identified by simple inspection. It has become common practice to include several interfaces on a single card and simply select the one required.

SoundBlaster interface

The SoundBlaster interface is probably the most popular of all the CD-ROM interfaces just because many CD-ROM drives are sold as part of a multimedia upgrade that includes a sound card as well. As such, CD-ROM and sound card bundles are popular and have the added advantage of a known system. The interface is part of the SoundBlaster Pro and greater models and its interface is independent of the sound card itself.

Panasonic

This is the same interface as is used on the SoundBlaster Pro card. It is used to link with Panasonic CD-ROM drives via a 40 pin IDC type connector and ribbon cable.

Mitsumi

This is another 40 pin interface which is quite commonly used. As its name suggests, it is offered on Mitsumi drives.

LMS

This interface is associated with early Philips single speed CM205 drive. It uses a 16 pin IDC connector and cable. Some early Mediavision sound cards supported this interface.

Sony

This interface uses a 34 pin IDC connector and is used on Sony drives.

SCSI

SCSI CD-ROM drives are a little more expensive and often require an additional SCSI interface, but they do support up to seven drives including hard disks and tape drives. This expansion is far greater than available from any of the other interfaces and thus is ideal for systems that require multiple CD-ROMs, such as development systems that store compilers and software development kits on CD-ROM, which can then be accessible directly without having to swap discs. Most of the high performance drives (triple and quad speed) have SCSI interfaces and, if state of the art performance is needed, the SCSI interface is probably the best choice.

Most SCSI adapters for the PC supply drivers that support the Microsoft MSCDEX CD-ROM standard for many of the standard SCSI drives available. In addition, the preferred interface for Windows NT installations is the SCSI interface, although there is support for SoundBlaster and Panasonic interfaces and they can be found by installing the Panasonic CD-ROM driver from the list of SCSI drivers supported by Windows NT. Although these interfaces do not use SCSI and are thus not logically expected to be located in the list of SCSI drivers, this is where they are found!

IDE

CD-ROM drives are now available with IDE interfaces that connect as the second drive to an IDE interface. Special drivers are needed to cope with the large data sizes and with the fact that CD-ROMs are removable and can be swapped out. Older IDE interface cards may not support IDE CD-ROMs and may need replacing with a newer Enhanced IDE card. As these cards are so cheap today, it is almost worthwhile getting a new IDE card just in case — thus saving the need to reorder or go back to the shop!

ISO 9660

Users of early CD-ROM releases often experienced problems with file formats as a result of a failure to define the higher level formatting needed. Although the low level structures, such as the sector structure, are defined within the Yellow Book specification, it was not until an industry group of companies got together to define the High Sierra format that CD-ROMs became interchangeable between platforms.

This de facto standard was adopted by the International Standards Organisation (ISO) in 1988 and has become the ISO 9660 standard — the predominant standard used today. This is not the only file format standard used as other operating system specific formats have also been developed and used. Such discs are normally not interchangeable between platforms because of the addition of different structures. A good example is the Apple Macintosh HFS format, which has a file structure that has two components called the resource and data forks. This is not supported within an MS-DOS FAT file format and is thus not compatible with an MS-DOS or Windows system. Other incompatibilities are caused by file naming conventions and the size of file names and directories.

Kodak's PhotoCD

Kodak's PhotoCD technology is a method of storing up to 100 compressed photographic images on a single disc. The images are stored at different resolutions, allowing the user to select the appropriate resolution for whatever use is needed. The file format used is the same as that for CD-i and CD-ROM XA. Typically, the images are transferred by a photography laboratory using very high quality scanners.

Photo CD discs are readable on CD-ROM XA and CD-i devices with the appropriate software. As standard, the discs are supplied with viewers for both IBM PC and Apple MAC computers. They are not normally readable on a standard CD-ROM drive, although some special drivers can read such discs using these drives, albeit with some limitations — usually only the first recorded session is available. Nearly all the CD-ROM drives from about 1991 onward are fully PhotoCD compatible and support multiple sessions. Kodak also offer their own stand-alone PhotoCD reader, which displays the pictures on a domestic television, as well as playing back audio CDs. In addition, most CD-i players can also playback PhotoCD images.

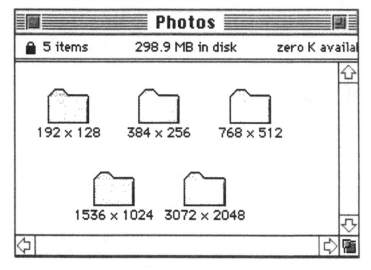

PhotoCD supported resolutions

PhotoCD viewer for Apple MAC

This technology is quite exciting because of its ability to provide high quality digitised images without having to have access to high quality scanners. The photos used in the diagrams in this book were extracted from PhotoCD and imported in Color-It or PhotoShop for manipulation.

Recordable CD-ROMs

CD-R technology, as defined in the Orange Book specification, defines the method of adding data to a CD-ROM disc and thus defines the method of overcoming the read only restriction that CDs have. With the ability to add data in a defined method to create what is known as a multi-session disc, all that was needed was a technique to overcome the mastering process to create a low cost method of creating small runs — even individual CDs — without the expense of producing a master and pressing the duplicate discs.

The problem was in providing a technique that allowed the pits to be created. The method used today involves a disc that is coated with a special dye. A CD-ROM writer is very similar to a reader, except that it has the ability to send a stronger laser beam to the disc which is powerful enough and matches a specific wavelength to alter the physical characteristic of the die at a specific location. When the beam is sent and lands on the dye, the dye is changed so that it will disperse the reading laser beam and thus, the equivalent of a pit is created. Unmodified dye represents the land area.

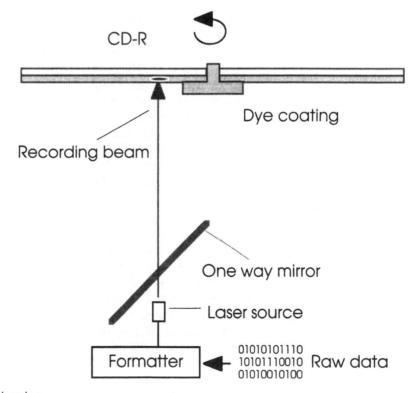

CD-R recording mechanics

This technology has provided a low cost method of archiving data and distributing software — especially alpha and beta releases, where the expense of committing to a normal master and press run should be avoided. CD-R discs that are written to once to create a single session disc are compatible with all CD-ROM drives. However, multi-session discs (where data has been appended) are not supported by older drives i.e. those before about 1990 or 1991. However, the technology is not all sweetness and light.

CD-ROM writers are still expensive, compared with read-only devices. Typically, this is still a factor of five — despite the cost of writers coming down by a factor of four over the last two years. The process is slow: the writer transfers data at the same speed that is used to read the data and at 300 kb per second for a double speed drive, transferring 600 Mbytes is not a fast process. In addition, a fairly powerful PC is needed, with a large amount of free disk space. The CD is normally simulated using disk space and then is recorded from that disk. This means that a 750 Mbyte PC disk drive is needed. This drive must also be very fast, to ensure that there are no intermissions in the recording process, although the need is decreasing by the use of data caches within the recorder. However, once the disc is created, the possibilities are unlimited.

CD-i

Compact Disc Interactive or CD-i is not so much a disc standard but a definition for an integrated system that can deliver interactive data, video and audio to a television or stereo system without the need for a computer. The system was defined in 1987 by Philips and Sony and is based around Motorola's M68000 processor technology. The specification is often referred to as the Green Book. CD-i uses Microware's OS-9 real time operating system internally, although this is completely embedded and hidden from the user.

Based on the CD-ROM basic format, CD-i allows different sector types to be defined which support interleaved audio and video for example. The integrated system can extract this data, decode it, display it and simultaneously access the next data streams. This capability provides the user with a seamless multimedia system as a consumer item that is simply hooked into the domestic television. The systems playback audio CDs and also access PhotoCDs.

CD-i material is stored on a CD-ROM which contains all the interactive material. The system can be used for games, electronic references (such as encyclopaedias), movies and anything that uses audio, video or data.

The audio part of the system provides six audio options, comprising three quality levels in either mono or stereo:

- A-Level is equivalent to a high quality audio LP with a total playing time of 2 hours.

- B-Level is equivalent to the best FM broadcasts.
- C-Level is equivalent to the best AM broadcasts with a total playing time of 16 hours.
- MPEG audio is also supported.

Video support is based on four video planes to allow composite images to be created. There is a fixed background, two full screen planes (to allow two separate video sources to be played back simultaneously) and a small cursor in the foreground to allow the user to interact with the system. The system also supports special effects such as dissolving images, cutting and wiping. MPEG1 video is also included and it is possible to play MPEG1 movies from CD-i discs. Three video quality levels are supported:

- Normal resolution for most video
- Double resolution for text and graphics
- High resolution for HDTV

DVI

This technology was bought from General Electric by Intel in 1987, one year after its launch in 1986. It is similar to CD-i but is based around the Intel 80x86 PC architecture for the controller. It also supports synchronised and interactive video, audio and data but uses proprietary technology which has been overtaken by open compression standards, such as MPEG.

DVI is capable of supporting full motion video running at 30 frames per second with a frame size of 512 by 482 pixels. The compression technology allows a 72 minute movie to be stored on a single disc.

CD-ROM XA

This standard was defined in 1988 by Philips, Sony and Microsoft as Microsoft's standard for multimedia applications. It is similar to CD-i in that it supports interleaved and synchronised audio and video but it does not rely on an M68000 based system. It uses the ISO 9660 file format and the mode 1 error corrected sector format as well as some new formats specified through additional header information.

The original specification did not support video compression, although audio compression using ADPCM was used. This required additional hardware support and, despite a lot of initial interest in the standard, it has faded into the background in recent years.

Digital Video Disc

Digital Video Disc or DVD is a development of the CD-ROM that dramatically increases its storage capacity and thus makes it possible to store complete movies with high quality soundtracks, including alternative soundtracks for multi-lingual support. It is interesting to note that the CDi system was also designed to provide a suitable medium for movies but did not gain sufficient market acceptance. One possible reason is that a full length feature film required two discs which meant that the file was interrupted while the discs were swapped over. The DVD system uses a combination of a new CD-ROM technology and MPEG-2 video and audio compression to provide a new method of delivering digital video.

The DVD disc format

A DVD is similar to that of a CD-ROM in that the data is stored by creating tiny pits in a spiral on the disc. A laser beam is again used to detect where the pits are located and where they are not and this information is decoded to obtain the digital data. A DVD disc is the same size (120 mm) as the CD-ROM and many players are backwardly compatible with CD-ROMs. It is rapidly becoming established as the latest CD-ROM standard for personal computers. DVD increases the storage capacity by using several methods:

- The size of the pits is reduced and the spiral tightened so that more data can be physically encoded on the disc. To recognise these smaller structures, a smaller wavelength laser is used along with fine resolution optics to move the beam as it scans over the disc.

  The spin speed of the single layer disc is constant and set to 3.49 meters per second, which is slower than current multispeed CD-ROM drives. The overall data read rate is 11.08 Mb/s which delivers audio/video media payload of 9.8 Mb/s. That is equivalent to an 8X CD-ROM.

- It uses a second layer to double the storage on a single disc. The second layer is accessed by changing the laser beam focus so that it is focused onto the second layer pits instead of the first layer structures.

- Unlike a CD-ROM, both sides of the disc can be used to further increase the data storage. This means that a two-layer, two-sided disc can store 17 billion bytes compared to the normal 4.7 billion bytes that a single-sided, single-layer DVD can hold.

  A double sided disc is made by bonding two .6 mm thick substrates together to form a 1.2 mm to 1.5 mm thick disc with data in the middle, and the same physical size as a CD-ROM. The single sided DVD can store 135 minutes of video which is enough for a feature length film and it is likely that this format will be the most used. The increased capacity offered by the other format types offers some other interesting possibilities. It is possible to contain multiple video streams so that different camera angles can be selected of the same event. This is attractive for sports programs where the user can then interact with the DVD system and select different viewpoints as required. For films, it means that complete trilogies can be included if needed.

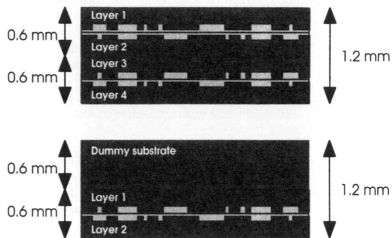

*A four layer and a two layer
DVD disc in cross section*

	Diameter	Thickness	Track pitch	Minimum pit length	Laser wavelength	Data capacity (per layer)	Layers
DVD	120 mm	0.6 mm	0.74 nm	0.40 nm	640 nm	4.7 GB	1,2,4
CD	120 mm	1.2 mm	1.6 nm	0.834 nm	780 nm	.68 GB	1

DVD and CD specifications

Video encoding

DVD-Video uses a restricted subset of MPEG-2 fixed or variable bit rate video, allowing one channel of video in an ISO 13818-1 (or ITU H.222) Program Stream with 4:3 or 16:9 aspect ratios. To support normal size screens, letterbox filters are built into the player that shrink the height of the picture to 75% so that 16:9 aspect ratio pictures can be shown full width on a standard 4:3 TV screen. There is also a 'pan and scan' mode which shows a portion of the 16:9 image at full height on a 4:3 screen. This is achieved by using 'centre of interest' co-ordinates that are encoded in the video stream to select the part of the frame that will be shown. This is similar to the technique that is used currently to compensate for the wider format that many films were originally shot in. The disadvantage is that peripheral action and interest are lost and where there may be several scenes of equal importance, one scene may have to be chosen in preference to the other.

The MPEG-2 video encoding used is the 'Main Profile at Main Level' along with variable bit rate encoding that adapts the bit rate allocation depending on the complexity of the video material itself. This means that where less bits are needed for a given quality, they can be utilised elsewhere in the stream to improve/maintain quality.

The video material is described and marked in terms of content and this provides the basic information for parental control and restricted access for children as well as providing support for interactive modes. This can be based on a frame-by-frame basis as well as by program. Combining this with different edited versions of the same film would allow different audiences to see an appropriate version or even a film with alternative endings!

The disc also provides a better mechanism for slow motion playback and random searching which again increases its potential for use with interactive material.

Audio

DVD-Audio supports up to eight streams of MPEG-2 audio in either 2.0 channel, 5.1 channel, or 7.1 channel audio or Dolby AC-3 5.1 channel audio. A nice feature of MPEG-2 multichannel audio is that it is downward compatible. For instance, an MPEG-1 decoder will be able to play the stereo portion of a 7.1 channel stream and a 5.1 channel MPEG-2 decoder will be able to play six of the eight channels. The stereo portion can be encoded with Dolby ProLogic so that it will work with existing surround sound systems. With AC-3, the 5.1 channel signals must be decoded to six channels and mixed together to create a stereo signal. Players are only required to support stereo output, so some will require external AC-3 or MPEG-2 multichannel decoders. IEC-958 serial digital audio distribution is the likely method of interconnection.

Sub-pictures

Sub-pictures as described in the DVD standards are bit map streams multiplexed with the audio/video program stream. They are used to provide graphics information for use in menus, subtitles and programme credits. Each of the 32 separate sub-picture channels defined can be up to 720 x 480 x 4 colours in size and can be controlled using a simple set of display commands. They also provide additional information for use in interactive DVD material.

A replacement for VHS?

Both the video and audio quality are better than can be achieved with current broadcast technology and VHS tape formats, although some suggest that the MPEG2 compression introduces artefacts of its own. The audio is compressed using MPEG-2 audio or Dolby AC3 Surround Sound. If AC3 is used, a separate ProLogic Surround Sound decoder can be used to provide surround sound if required. Both the audio, video and information data streams are multiplexed together to create the single data stream that is encoded onto the DVD disc.

DVD is poised to replace VHS tape as the preferred distribution medium. It is about 25% of the cost of a tape and can hold substantially more content than a tape. It is a more robust medium and through its interactive nature can potentially offer more — or a higher quality — content than currently available. As a result, it is being backed by the content providers such as the Hollywood film studios as an ideal way of bringing their product to market. It also bridges the gap between personal computers and the more consumer orientated television and video recorder and this may be the catalyst that finally brings these technologies together. The only stumbling block for this technology concerns copyright protection and the fear of the content providers that the material will simply be copied and illegally distributed. While this does occur with analogue tape, each copy degrades and therefore the quality can be extremely poor. With the content digitally encoded, each copy is an exact replica of the original and this has created a concern and a requirement for copy-protection mechanisms. This problem will no doubt be addressed at some point.

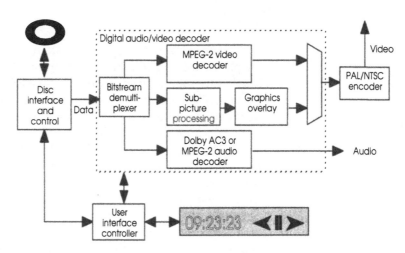

A DVD player with video/audio decoder

VHS tape still has the advantage of being a read / write medium and will probably still have a place because of this and there is a very large number of VHS recorders and players in the market already. However, DVD could well establish itself as the future medium for consumer video in the same way that VHS recorders established the video rental business.

7   POTS/PSTN

The analogue telephone is the primary communication tool in use today and its use with modems has provided the main way of linking computers together. The explosion in use of the Internet via modems to local service providers demonstrates the power behind linking and communicating with computers. As a result, the division between the telephone, the FAX machine and the computer is becoming smaller every day. To understand how the telephone system works and how it is being used by computers, this chapter provides a tutorial on the technology behind the telephone.

POTS and PSTN

The plain old telephone system (POTS) or public switched telephone network (PSTN) refers to the analogue telephone. The term POTS is frequently used in the US whilst PSTN is used in the UK and Europe. This technology, originally based on mechanical components, uses a combination of switching voltages for control on the same two lines that carry the audio signals.

The telephone line itself is a twisted pair cable that creates a loop between subscribers' telephones and the exchange. Inserted in this loop is a battery providing a 48 volt DC (50 volts in the UK) across the twisted pair with the positive connection on the tip wire and the negative connection on the ring wire. This nomenclature comes from the use of jack plugs in a manual exchange, where lines were physically exchanged by people to create a connection. The jack plug makes contact with its socket through the tip and its body or ring. The voltage across the twisted pair is used to signal to the exchange and to carry out tasks such as pulse dialling and ringing.

The battery is normally stored at the telephone exchange and is often a lead-acid battery which provides an independent supply of pure DC (without any AC ripple) in case of power cuts. This also allows electronic telephones to draw current while in use and thus not need to have batteries or an external power supply.

When the telephone is not in use, it has a high resistance so that the current drawn is only a few micro-amps. This is a mandatory requirement so that the telephone company does not provide a free source of power to the subscriber. When the telephone is used and goes on hook, its resistance across the loop decreases and the voltage drops to about 3 to 9 volts drawing a current of not less than 20 milliamps. In practice, the actual values depend on the length of the loop and hence its resistance. The 20 milliamp figure determines how much power a telephone can draw. If it needs more power, this must be supplied locally. For most telephones with simple facilities, the on-line power is sufficient. For others, such as cordless or those with built in answer machines, additional local power is needed.

The audio input and output are converted from a four wire system with separate grounds and signal paths to a two wire connection that can use the tip and ring connections provided by the twisted pair. This conversion used to be done using capacitor-choke networks to allow the audio to be superimposed on each line. Today, it is done electronically.

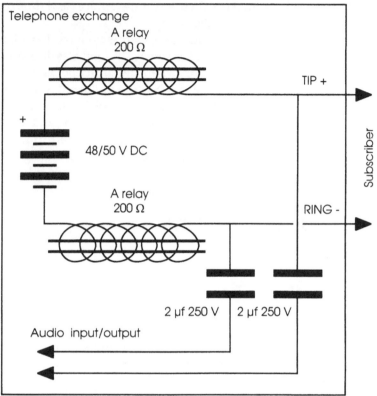

Analogue telephone subscriber/
exchange diagram

At the telephone exchange side, as shown in the diagram, are the components needed to interface to the subscriber. The tip and ring connections come in to the exchange and are connected to the battery via two chokes, which are actually the coils of a relay. The relay, often known as the A relay, is energised when the subscriber telephone goes off-hook and starts to draw current. This changes contacts on the relay to signal to the exchange that the telephone is off-hook. By using these contacts, different audio sources, such as the dial tone and the other party, can be connected in turn. The relay coils also block any AC transmission between the battery and the subscriber so that the battery only sees a DC world and not the audio signals as well. The two capacitors are used to couple the audio signals from the loop to the exchange.

In addition to this, there are protection circuits that protect the line from over voltages, as can be caused by power spikes and lightning strikes.

Within the handset itself, there is a microphone and speaker to support the two way audio. They are connected so that some of the signal from the microphone is heard in the earpiece speaker. This makes the telephone feel more alive and not acoustically dead and also provides some aural feedback to stop people shouting down the telephone! This audio signal is known as sidetone. Too much sidetone makes callers lower their voices and thus reduces their intelligibility to the other party.

In addition, there is circuitry that adjusts the earpiece output so that it is not too loud. The loudness depends on the length of the loop. A short loop, where the subscriber is close to the exchange, will have a louder signal because of the reduced drop across the twisted pair cable. To compensate, a voltage sensitive resistor used to be used to maintain a consistent loop loading and thus audio signal. Today, this loop compensation function is performed electronically.

The bandwidth used on an analogue telephone is limited to the range 300 Hz to 3.1 kHz, giving a bandwidth of about 3 kHz. Audio outside of this range is lost.

Dialling and ringing

Two types of dialling technique are used: pulse dialling and touch-tone.

Pulse dialling

Pulse dialling is based on the mechanical rotary dial and is essentially the oldest form of dialling. Rotating the dial back and letting it go generates a number of pulses which can be counted by the exchange to work out the number. The pulses are transmitted by disconnecting and connecting the phone i.e. making it go on hook and off hook. This causes one of the A relays to pulse and, if connected to a counter, to calculate the number. Electro-mechanical exchanges use these counters to make the connections as each number comes in.

This electro-mechanical method is now performed electronically. The advantage that this gives, apart from the removal of the rotary dial and the ability to support push buttons, is that the dialler chip or function can maintain a list of numbers, including the last number dialled, and make them available through special buttons or key sequences.

Encoding for the pulse dialling can be heard through the sidetone as clicks. Normally, the digits 1 to 9 are encoded by the same number of pulses i.e. a one is a single pulse and 9 is nine pulses. The zero is encoded as ten pulses. The pulse rate is typically 6 to 15 pulses per second and was determined by the speed of the electro-mechanical equipment in the exchanges that used the pulses to make the connection.

The pulses have to conform to a specified mark/space ratio. In the US this is 60/40 while in Europe, the ratio is 63/37.

Touch-tone

Touch-tone dialling or DTMF uses special pairs of tones to indicate numbers or special keys. They have a big advantage over the pulse method in that they cannot only be used to dial a number without having to wait for the pulsing, but they can also be used to provide control signals and other information during a call.

Each tone is sent continuously while the corresponding key is pressed down. The minimum burst of tone that can be decoded is 100 ms and this can lead to fast automatic dialling by feature telephones and PCs.

Digit/key	Tone pairs	
0	941	1336
1	697	1209
2	697	1336
3	697	1477
4	770	1209
5	770	1336
6	770	1477
7	852	1209
8	852	1336
9	852	1477
*	942	1209
B	697	1633
C	770	1633
D	852	1633
#	941	1477
F	941	1633

Note: Keys B, C, D and F are spare and not defined

DTMF tone encoding

To support DTMF, both the telephone and the telephone exchange need to be able to generate and decode the tones. This is done electronically by special circuits.

The ringer

The ringer is an important part of a telephone and is the audible device that rings, warbles or flashes when an incoming call is present. The ringing is actually generated by the telephone exchange sending an AC waveform down through the line at a frequency between 15 and 68 Hz, although 20 Hz is popular in the US. The voltage can be quite high and is between 48 and 150 volts, depending on the length of the loop and the number of ringers attached.

The ringer itself consists of a resonant circuit that is placed across the loop when the telephone is on-hook. When the AC voltage is placed on the line, the ringer activates because of the resonant match and makes a noise. Like most aspects of a telephone, this ringer is electronic and normally generates a signal from the ring waveform and outputs it via a small speaker or piezo-electric transducer.

The ringer has an important use for the telephone company because it is connected across the loop while the telephone is not in use. The more telephones that are connected, the higher the loading. This can be seen by the telephone company and in this way they can tell if new equipment has been added.

There is only a certain amount of energy within the waveform that can be sent down the line. If there are too many ringers connected across the loop (i.e. to many telephones), there will be a point when none of the ringers will work. This is against the specifications, and telephones must be removed until they all work. This has led to the development and use of the ring equivalent number that can be used to calculate how many telephones can be connected. If the subscriber supports a REN of 4.0, four telephones, each having a REN of 1.0 can be fitted. Alternatively, two telephones with a REN of 1.5 each and two telephones with a REN of 0.5 can be used.

The exchange or switch

Until now, the discussion has centred on the subscriber end view of what the exchange provided. The development of the exchange is fundamental to the continuing goal of being able to provide high speed data communications as easily as it is today to speak to someone anywhere in the world.

The exchange (or switch, as it is also known) has three main functions to perform: it provides a consistent interface to the subscriber, it routes the various calls to their destinations and it provides billing information to the telecommunications company (telco). Up until the introduction of electronics, the two twisted pair cable has severely limited the ability to provide more control and signalling information necessary to improve and expand the service. The limitations imposed on the subscriber interface by the original technology meant that the exchanges had to become more intelligent, enabling signalling and control to be incorporated. As a result, the interface between the subscriber and exchange has remained physically the same, except that today there is far more interaction between the two.

The old mechanical exchanges could not support the intelligence required and it was not until the introduction of intelligent electronic interface circuits that the telephone network really started to advance. Although electronics had been introduced into exchanges to replace the old relay and rotary switches originally used to make the connections, this was limited to the introduction of routing circuits such as cross point switches.

The advent of the digital and intelligent exchange allowed not only the initial mechanical components to be replaced but also allowed support for more sophisticated processes, such as dialling using DTMF.

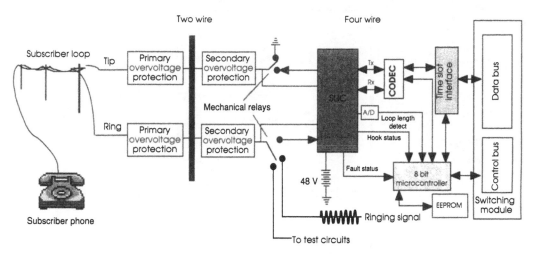

Digital subscriber line card

With a digital telephone system, the subscriber's telephone is connected to a line card which provides the same interface but uses digital signalling to connect the call and converts the analogue speech into a digital bitstream using a codec. Dialling involves sending messages into the public network to create the links necessary to connect the call. Once the connection is made, the speech is transferred between the two parties. The data is inserted into a time slot in a larger bitstream containing multiple calls and placed into the public network, which consists of very high bandwidth links between multiple exchanges. In this way, the overall public network becomes a nebulous, amorphous mass, where the exchange asks for connections and receives them, and data is inserted and 'magically' appears at the other end.

A block diagram of a digital line card is shown. The battery, ringer wave form and so on are still there, together with the over-voltage protection. However, the SLIC (subscriber line interface chip) device now performs the two to four wire conversion, detects the pulse dialling, decodes any DTMF tones and so on. The analogue speech is converted by the codec into an eight bit PCM encoded digital stream. The SLIC also provides status information about the loop length and the on hook and fault status. This information is passed to a small micro-controller which provides the intelligence within the card. The digital data and control messages are sent to a switching module using a time slot within the data stream that is used to communicate within that module. This is done by a combination of the time slot interface and micro-controller.

When a telephone number is dialled, it is collated by the micro-controller which sends it as a request to create the call connection to the switch. Once the connection has been made, the digital data is routed through the switch to its destination.

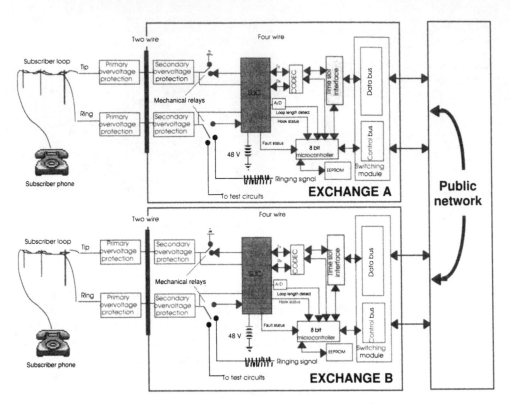

Connecting two remote callers

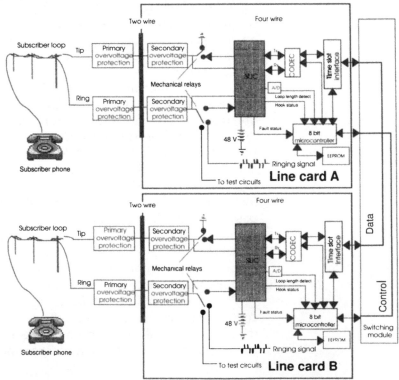

Local call connection

The micro-controller can interpret the information from the SLIC and thus improve its intelligence. It can switch the connection to a test circuit. It can also send a warning tone if the telephone has been left off hook for some time to warn the subscriber that the telephone has been left off hook by mistake.

If a call's destination is local or within the exchange, it can be connected directly and, in this case, the connection is from one line card to another. If the connection goes outside the exchange, the messages are sent to the public network. This network consists of a communication network between all the exchanges and effectively creates a large switch. Typically, a hierarchical architecture is used which tries to connect the call at the lowest possible level. If this is not possible, the call is presented at the next level, and so on, until it either reaches the highest point or the call connection is made. All this is done using digital messaging. By adding more information, additional subscriber services can be provided, such as call line identification, answer services, call waiting, and so on.

Providing digital communication

Even with the advent of digital exchanges and the potential they offer, the computer is still stuck with the analogue telephone interface as the means of getting connected into the network and accessing other users. Whilst LANs are fine for small local groups within an office, the distances involved in connecting computers across the country or even from one side of a city to another means that the telephone network is still the only really viable method.

The modem was developed to address this problem. It is essentially a telephone which takes a digital bitstream and encodes it using audio signals which can be sent down a telephone wire to another modem. Whilst modem technology has improved, and the current 28.8 kbaud modems achieving data transfer rates with data compression of over 100 kbits per second, this is still very slow when compared to a LAN or where the bandwidth required to transfer multimedia data is concerned.

As a result, the telcos have responded by changing the subscriber interface and providing an entirely digital interface which is ideal for the computer. This interface known as ISDN (integrated services digital network) can provide a far higher bandwidth than is available with an analogue telephone line. A second approach has been to define standards that can increase the bandwidth which can be obtained from a twisted pair link or fibre optic, so that in the future, they can deliver sufficient bandwidth to the user to support the multimedia applications that will appear, such as video servers and video on demand. The next four chapters will look at these technologies.

8   ISDN

ISDN (integrated services digital network) is a development from IDN (integrated digital network) that has performed the transmission and switching of phone calls within digital exchanges and switches. It is a standard that has been around for a long time and but has only recently been gaining popularity. Introduced in the mid 1980s, its provision to the subscriber was dependent on digital ISDN compatible switches being in place plus sufficient bandwidth being available to support them. As a result, the introduction of ISDN services in many countries depended on the replacement of non-compatible exchanges and switches and this has taken many years. It must also be said that the line tariffs and cost of adapter cards have slowed its introduction but this issue is also being addressed.

The International Telecommunications Union define ISDN as:

'A Network evolved from the telephony Integrated Digital Network (IDN) that provides end to end connectivity to support a wide range of services, including voice and non-voice service, to which users have access by a standard multipurpose user-network interface.'

The ISDN specification (I.411) describes all the characteristics that can be seen by the subscriber, including the signalling protocols, services and physical attributes of an ISDN connection. Despite this, many implementations in different countries are slight variants of this causing compatibility problems. The wider use of ISDN, particularly with video conferencing across international boundaries, has resulted in a lot of work to resolve the differences and thus allow a seamless connection across the world.

ISDN uses a double twisted pair cable but instead of using AC and DC voltages to provide the signalling protocol, the signalling is performed digitally by sending and receiving messages down a special channel — the D channel. Actual user data is sent using a B channel at speeds up to 64 kbits per second and two B channels are supported on a single twisted pair cable. This 2B + D configuration is known as a basic rate interface. Each B channel is normally treated as a separate call and thus using the two B channels together to double the available bandwidth doubles the call tariff.

The B channels are used for different forms of communication as defined by the bearer mode. The following modes are available:

Voice   Regular 3.1 kHz analogue voice service. Bit integrity is not assured. This is normally used for a telephone call to another audio only telephone. The bit integrity aspect is important as it effectively excludes data or fax transmission or the use of a modem.

Speech   G.711 speech transmission on the call. This is the equivalent of a normal analogue telephone call and does allow the use of a modem or fax.

Multi-use As defined by ISDN.

Data   Unrestricted data transfer. The data rate is specified separately. This is the call mode that is used for an H.320 call.

Alternate speech and data

The alternate transfer of speech and unrestricted data on a call (ISDN).

Non call-associated signalling

This provides a clear signalling path from the application to the service provider.

Of these modes, the most popular are the voice, speech and data modes.

ISDN reference model

The ISDN system consists of several interfaces and options. It is described using the reference model shown in the diagram. At the subscriber's end, ISDN is made available using the S interface, which allows ISDN equipment, known as a TE (terminal equipment) to be plugged into the ISDN socket. Non-ISDN compatible equipment can be connected via a terminal adapter (TA). Beyond the S interface is the NT2 block.

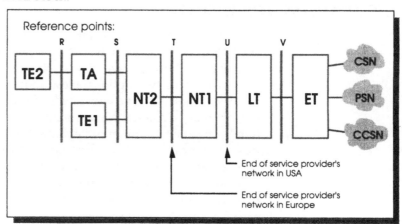

CSN:	Circuit switched network
PSN:	Packet switched network
CCSN:	Common channel signalling network
ET:	Exchange termination (C.O. switch)
LT:	Line termination (Line card)
NT1:	Network termination 1 (OSI layer 1 only)
NT2:	Network termination 2 (OSI layers 2 & 3)
TE1:	Terminal equipment 1 (ISDN terminal)
TA:	Terminal adapter
TE2:	Terminal equipment 2 (non-ISDN terminal)

ISDN reference model

This converts the S interface information and collates it into a higher performance and higher bandwidth bitstream where multiple ISDN calls are transferred to the next block, the NT1. The NT1 block provides support for the next two OSI model layers before sending the data to the line card (LT) using the U interface. From there, the data is sent into the switch and thence to the public network for switching and routing.

National differences in the model exist. In Europe, the service provider takes responsibility for the upper layers of the network as far as the T interface. In the US, the responsibility is as far as the U interface. This is of importance for ISDN switchboards and determines their connection point into the network.

ISDN standards

ISDN is defined by the following set of standards:

I.100	Vocabulary
I.200	Service aspects
I.300	Network aspects
I.400	User-network interface
I.500	Network interfaces
I.600	Maintenance principles
Q.93x	D channel signalling

User-network interfaces

The user has a choice of three different interfaces at subscriber level. The most common is the point to point interface which gives the longest cabling option and allows a single terminal to be connected.

The other two options allow multiple terminals to be connected on the same cable with a reduced cabling length. This allows the terminals to share the line but also through multiple subscriber numbering allocate a unique telephone number for each TE. By doing this, a caller can directly access a specific terminal. This is useful to ensure that a fax call goes to the fax machine and so on. This ability to support multiple TE installations on the same cable does not mean that all the units can be used at the same time, although two calls using each B channel could be supported.

The subscriber interface

The physical interface consists of two twisted pair cables giving four conductors in total, with two used for transmission and two for receiving data. The data capacity is based on a channelled approach with two B channels supporting 64 kbits per second and a third D channel that provides 16 kbits per second. The two B channels are bi-directional and used for data transfer. The D channel is primarily used for control and signalling, although it can be used for data transfer as well in some cases. The D channel uses a special higher level protocol called LAPD that is based on the HDLC frames.

All three channels are contained in a 48 bit frame running at 4 kHz. This gives a 192 kbits per second bandwidth carrying 144 useful bits (B1+B2+D) + 48 service bits within each 4 by 48 bit group. The service bits are used for frame alignment, collision detection and other functions.

Point to point interface

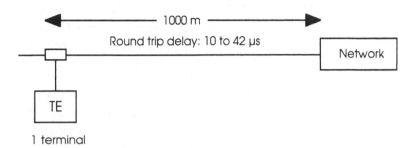

Passive bus interface

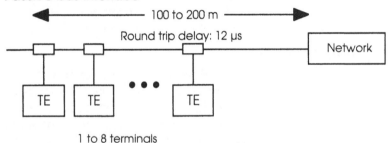

Extended bus interface

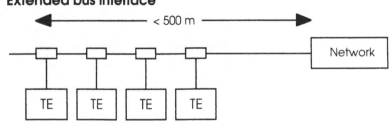

ISDN user-network interfaces

AMI coding

The data is encoded on the wires using a pseudo-ternary code or alternate mark inversion. With this technique, a zero is coded by being either a positive or a negative transition. The mid point value indicates a one. The encoding scheme is shown in the diagram. For every change from a one to a zero, a zero to one or from a zero to zero, the signal moves from one level to another. Only with a set of consecutive ones does it remain constant.

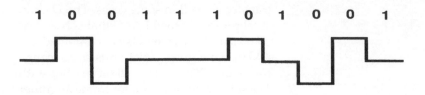

AMI encoding

Frame structure

The ISDN frame structure for a basic rate 48 bit frame is shown in the diagram. Frames are sent in both directions with the frame from the network to the terminal, carrying an E (echo) bit based on the contents of the last set of D bits that the terminal sent to it. Both frames are synchronised and this allows the E bit to be used to detect multiple simultaneous access by multiple terminals — more about this later. Terminal equipment that is sitting on the ISDN connection receives the frames from the network, extracts the information that is pertinent to them and ignores the rest. The D channel information that is extracted uses the LAPD protocol for its information and this also has to be decoded so that any relevant messages can be read. The service bits within the frame perform several housekeeping functions.

Synchronisation

When a terminal is connected to the network and activated, it synchronises with the network at several levels: bit, frame and multiple frames. The bit clock is recovered using the transitions within the AMI coding. The frame and multi-frame clocks are obtained from the F/F and Fa/N bits within the frame.

Activation

Layer 1 of the ISDN protocol specifies an activation and deactivation procedure which allows a terminal to be deactivated and thus go into a low power mode to conserve power. Activation and deactivation requests are conveyed across the interface by using the following digital signals called info signals. Three signals are specified. They sometimes use the framing structure, depending on the messaging requirements.

Info 0   If there is a no line signal, this indicates a deactivation request from the network to the terminal equipment and/or confirmation to the network that a terminal has deactivated.

Info 1   This is a signal that is not synchronised to the network timing. When it is transmitted from the terminal to the network, it is interpreted by the network as a request to activate the connection. It is used when the terminal wants to make a call and has come out of a deactivated state.

Info 2   This message consists of layer 1 frames with a high density of 0s to allow fast synchronisation of terminals. Zeros cause transitions on the line and thus allow the bit clock to be quickly extracted. It is used by the network to activate the terminal or as confirmation that the network has been activated in response to Info 1.

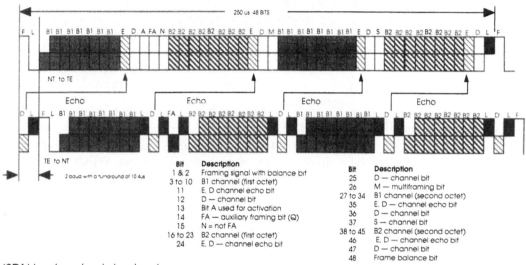

ISDN basic rate data structure

Bit	Description		Bit	Description
1 & 2	Framing signal with balance bit		25	D — channel bit
3 to 10	B1 channel (first octet)		26	M — multiframing bit
11	E, D channel echo bit		27 to 34	B1 channel (second octet)
12	D — channel bit		35	E, D — channel echo bit
13	Bit A used for activation		36	D — channel bit
14	FA — auxiliary framing bit (Q)		37	S — channel bit
15	N = not FA		38 to 45	B2 channel (second octet)
16 to 23	B2 channel (first octet)		46	E, D — channel echo bit
24	E, D — channel echo bit		47	D — channel bit
			48	Frame balance bit

S/T multiframing

Within the frame is a multiframe bit. This can be used to provide a low speed signalling channel in addition to the D channel. The channel is split into five sub-channels (SC1 to SC5) when the network is communicating to the terminal equipment end. In the opposite direction, it is known as the Q channel. It provides a local channel for maintenance between the terminal and the network interface and can be used to signal problems such as U interface connection faults.

Multipoint support

With a point to point configuration with only a single terminal adapter, there is no problem with collisions. With multiple terminal adapters, there can be a problem when two or more terminals try to use the D channel at the same time. The technique used to resolve this is based around the CSMA/CR (Collision Sense Multiple Access, Collision Recognition) method. This allows terminals to detect if another device is transmitting and accessing the channel.

The technique works as follows:
- All the terminals implement the LAPD protocol on the D channel. This forces them to request permission to access the B channels via D channel messages.

- All terminals monitor the echo bit in the incoming frames from the network on the echo channel. This bit reflects the last received D channel bit from the terminal end of the connection.
- To avoid collision, a terminal examines the E echo channel to see if its last transmitted D bit matches the next available E echo bit. If these two are the same the TE continues, if not it stops transmission and returns to monitoring the E echo channel.
- During transmission, the echo bit is checked to ensure it matches. If not, a collision has occurred.
- Before accessing the D channel, a terminal must wait until eight successive 1s have been sent via the E bit.
- When monitoring, all terminals send a set of 1s which are logically anded with the accessing terminals data to create the data frame. This effectively allows the terminals to sit on the common bus.

Multiple subscriber numbering

This is a facility that allows individual terminals on an ISDN line to be allocated a unique telephone number so that a connection can be made direct to that terminal. This is useful for allowing a fax call to be directed to a fax machine, an H.320 video conferencing call to be accepted directly by a conferencing suite, and so on. The numbering schemes vary from country to country but typically comprise a generic number which identifies the ISDN subscriber with an extra digit to identify the particular terminal.

A further complication can occur with the basic rate interface: it can provide two B channels which can be used to support two simultaneous B channel calls. Some numbering schemes allocate a different number to each B channel, whilst others allocate the same number. With the former scheme, the number is not allocated physically to B channel 1 or 2 but is assigned to the next free channel, which may, of course, vary. The latter scheme works by allowing a second call to the same number to be made. The second call is allocated the free B channel.

A further complication with numbering is caused by the requirement to add another digit at the beginning of the number to ensure that a data channel is used and is not a speech line. This can cause problems if omitted and is frequently needed when making international calls.

D channel protocols

The D channel is used for messages and control information between the terminal and the network. The LAPD frame format used is shown in the diagram. It provides a wrapper around a block of data describing its destination, some control information and a CRC value for error correction.

Both the channel commands and status information that go between the terminal and the network are defined in the Q.93x specifications. For example, the channel is used to request a channel of a certain type to support a speech or data call. It can dial a number or, to be more accurate, request a connection to the subscriber's number. If the destination cannot support the call, the network signals this back. If the destination is an analogue number and the ISDN channel was requested for a voice call, the network performs the conversion itself and allows the call to proceed. This is not the case for a data call that incorporates voice data such as an H.320 call, however.

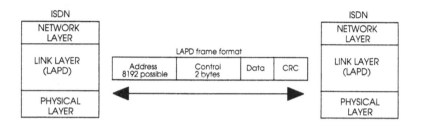

LAPD ISDN frame

Primary rate

So far, the main focus of this chapter has been on the basic rate interface but ISDN, through its I.431 specification, defines a higher bandwidth interface called primary rate. Primary rate provides 1.544 Mbits per second of bandwidth in the US implementation and 2.048 Mbits per second in the European implementation. It only supports the point to point configuration and does not allow multiple terminals to be directly connected.

The channel allocation within the interface is defined by the I.412 specification. It allocates the bandwidth in multiples of different size channels as follows:

- B channels with 64 kbits per second per channel
- H0 channels with 384 kbits per second per channel
- H1 channels with 1,536 kbits per second per channel
- D channels with 64 kbits per second per channel

The interface uses a basic frame with a period of 125 µs and a number of time slots, each containing 8 bits of information. The US standard defines 24 time slots to give the 1.544 Mbits per second total bandwidth. The European version uses 32 time slots with slot 0 allocated to frame alignment and slot 15 used for the D channel data.

ISDN configurations

Given the various components that make up an ISDN system, how are they fitted together and used?

The diagram shows two common configurations used for commercial applications. The top example shows the ISDN services being supplied by a PABX or switchboard from a direct link to the exchange of central office. In this configuration, the terminals are connected to the PABX as if they were individually connected to the ISDN services provided by a telco.

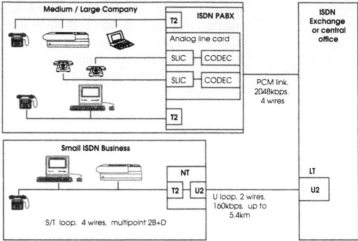

ISDN configurations (US model)

The difference is that the PABX can also cope with analogue telephones and equipment and through its own line cards, convert the analogue data into a digital format that can be combined to go across the PCM link to the exchange.

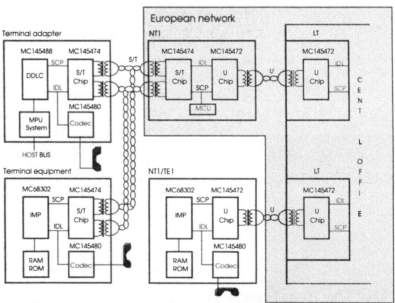

European ISDN network configuration

The second example shows the exchange providing a direct interface into a small business which is using a multipoint configuration to support multiple devices.Both these examples use the US model, where the telco provides a U type interface as the limit of its system and relies on the subscriber supplying the missing NT1 block to provide the more normal T interface. European configurations normally provide a T interface and effectively include the NT1 block within the network.

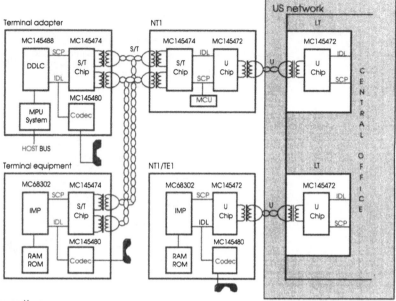

US ISDN network configuration

Making an ISDN call

The key to understanding how an ISDN call is made is to grasp how the various channels are used. All the data is transmitted within a frame, which is divided into four separate components:

- The housekeeping bits used by collision detection, synchronisation and so on.
- The data that forms the D channel.
- The data that goes into the first B channel.
- The data that goes into the second B channel.

The order is quite deliberate and reflects the order in which the components are used. The housekeeping bits are first used when the network sends its synchronising frames — the Info 2 message — in response to a terminal using the Info 1 message to activate the network. Once this has been done, the D channel is set up and available and becomes the method of sending and receiving messages. This provides a mechanism for requesting a B data channel, the bearer mode, and so on. Status information is sent back in reply from the network and this signalling process allows a call to be set up stage by stage. Once the call is connected, the B channel can be used for the specified data transmission.

The flow diagram shows a simplified flow for a terminal making a call using a single B channel. The collision monitoring is not included i.e. it assumes that the terminal is in a point to point configuration. The messaging has also been simplified to help bring out the main principles. The many time-outs which are involved to protect the interface if a terminal is disconnected or powered down, for example, are also not included.

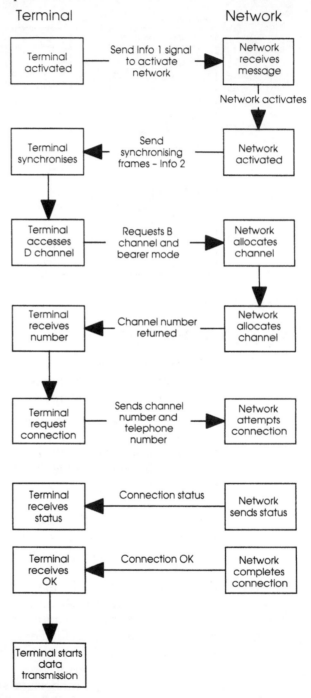

Simplified flow diagram for an ISDN call

The call starts by the terminal sending a request to the network to activate. The network in turn activates and starts sending Info 2 frames to allow the terminal to synchronise. In practice, it does this for a set period and if no further request is made, the network stops sending and deactivates. This procedure is the same as for the Info 0 message and the terminal should also deactivate.

Once synchronised, the D channel is used to start the process of making a call. The first action is to request a channel specifying the bearer mode e.g. voice, speech, data etc. If successful, the network replies with a channel number. The connection can then be made where the terminal supplies the number. During the connection phase, status information is returned by the network to the terminal. Eventually, 'connection made' messages are returned and the terminal can start to use the B channel to send data. If the connection is not successful, the terminal is informed of the problem via the D channel and it must decide what to do. Normally, it clears down the original channel.

Summary

ISDN is becoming more and more popular as the line tariffs and equipment reduce in cost and availability and interoperability improves. At the basic rate level, it provides a data rate far faster than can be achieved using a modem and an analogue line.

It provides a variety of bandwidths and, at the top end, is capable of supporting the ability to playback and MPEG1 encoded movies, for example. However, this may not be enough to meet the requirements that the industry is now projecting.

When ISDN was introduced, it was thought that the data rates would be sufficient for most computer users. However, its slow introduction and the rampant consumption of bandwidth within the computer industry, particularly with multimedia applications, has left it behind. It is still an efficient way of transferring data and, until the next generation telephone networks with ADSL and ATM are available, the only real universal digital communications method available today. Until there is an alternative, it cannot be ignored.

9 ADSL

In the never ending search for bandwidth to allow multimedia data to be delivered via a telephone network or down a cable, instead of from a local storage device such as a CD-ROM, ISDN does leave a little to be desired. Whilst it does provide more services and a higher bandwidth, when compared with analogue telephone line and modem technology, unless primary rate is used, the data rates are not sufficient to support MPEG1 quality video and audio, let alone MPEG2.

In addition, the cabling is different from that used by an analogue telephone and therefore someone — usually the subscriber — has to pay for it. As a result, although ISDN has a definite part to play in providing data communications and video conferencing, especially in the commercial marketplace, it is not the ideal solution for the consumer.

An ideal solution to address these problems would need to provide sufficient bandwidth for MPEG1 quality video and audio to be delivered to the home, support and be compatible with the analogue telephone system and, perhaps more importantly, the two wire twisted pair cabling that it uses. The provision of a back channel from the subscriber to the information source or provider to allow interactive support and to obtain instant billing information and payment, is also necessary to satisfy the financial aspects of such an enterprise.

Work to address these problems has culminated in the development of ADSL, which provides high bandwidth into the home using standard POTS cable.

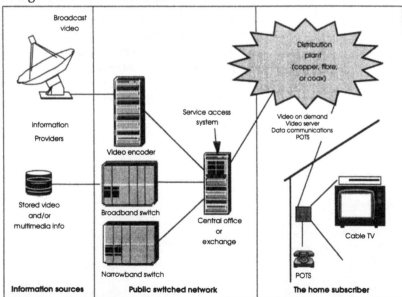

An ideal multimedia network

What is ADSL?

The ADSL — asymmetric digital subscriber line — specifications define how the existing analogue twisted pair cabling can be used to achieve far higher digital data rates than those offered by current ISDN implementations. Data transfer is asymmetric, in that the data bandwidth downstream from the network to the subscriber is far higher than that available upstream from the subscriber to the network. This asymmetric partitioning is to provide more bandwidth to support the transmission of MPEG encoded movies, whilst providing a fairly fast back channel to allow the subscriber to communicate with the information providers. The back channel also facilitates interactive participation, which is becoming a hallmark of many multimedia applications available today and planned for the future.

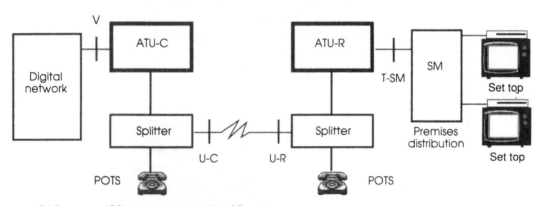

ATU-C:	ADSL transmission unit at the CO end.
ATU-R:	ADSL transmission unit at the remote end.
POTS:	Plain old telephone service
Splitter:	Filter which separates high and low frequency signals at the CO end and remote end.
SM:	Service module, performs terminal adaption functions. Supports either a bus or star based connection.
T-SM:	Interface(s) between ATU-R and SM(s).
U-R:	Loop interface at remote (customer) end.
U-C:	Loop interface at central office end.
V:	Logical interface between ATU-C and network element, such as one or more switching systems.

ADSL reference model

The reference model is shown in the previous diagram. It utilises the existing analogue twisted pair cable and analogue telephones. Instead of the telephone being directly connected to the line, it is connected via a splitter, which separates the existing low frequency POTS service from the higher frequency ADSL digital channels. This is done by filters. The splitter also combines the signals onto the line to support the two way communication and the ADSL upstream back channel. At the subscriber's end, the ADSL digital data is routed through a special transmission unit to a service module in the house that distributes the data to the appropriate terminals. In the diagram, the example is a digital television with a set top decoder.

A PC or other multimedia device could also be connected. The service module can support multiple units connected using a bus-based or star topology.

At the network end, the ADSL data and POTS connection are also connected by a splitter. To add ADSL support to an existing line-card is relatively simple as it only involves adding the digital data pathway and replacing/augmenting the line card with a splitter. This enables the service to be provided at a reasonable cost and without having to completely replace a switch.

Squeezing the bandwidth

The ability to obtain more digital data bandwidth from the existing cabling has been made possible by the provision through integrated circuits of low cost digital signal processors. These can not only encode large amounts of data, but also tolerate large amounts of noise and signal reflection through the use of sophisticated digital signal processing to recovery data from noisy signals.

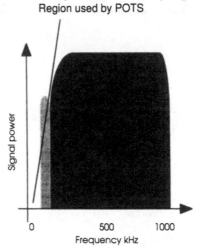

Region used by POTS

High bitrate ADSL downstream channel

Low bitrate ADSL upstream channel

Frequency allocation

ADSL data is encoded by modulating frequencies much higher than those used by a POTS service. A POTS telephone uses DC and very low frequency waveforms for its signalling and the highest frequency component is the top end of its supported audio bandwidth at 3.1 kHz. As the graph shows, this leaves a large amount of frequency bandwidth available to encode the upstream and downstream ADSL data.

Discrete multitone

The encoding technique which has been specified for ADSL encoding is the discrete multitone technique (DMT). This effectively divides the available bandwidth into channels which are modulated to encode the data. With the ADSL specification, 15 bits are specified per channel and this, multiplied by the modulation frequency, gives 60 kbits per second data transfer. By using a large number of channels, the high bandwidth can be obtained. The channels are asymmetrically allocated to give different bandwidths for the upstream and downstream channels.

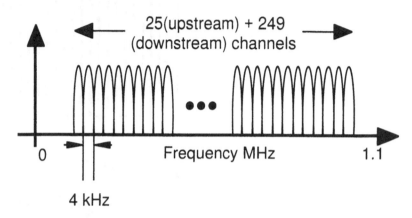

Data rates:
Each channel encodes 15 bits at 4 kHz =   60   kb/s
Downstream channel has 249 channels =   14.9 MB/s
Upstream channel has 249 channels =   1.5 MB/s

DMT channel distribution

The figures shown in the diagram represent a theoretical maximum and assume that the frequency response of the bandwidth is linear across all the frequencies. In practice, this is not the case and the bits per channel allocation is adjusted to compensate. This is shown in the next diagram.

The attenuation curve is based on the characteristics of the twisted pair cabling and associated components. This changes the bits per channel allocation with the frequency response. This explains why data rates are lower than those theoretically possible. However, this reduction to match the attenuation curve makes the data transmission far more reliable.

There is one further aspect to consider: the data rate is very dependent on the amount of noise included in the signal and how easily the tones can be extracted from the noise. The transmission rates are dependent on the signal processors that are built into the ADSL interface chips. The more noise, the less efficient the extraction and, therefore, the slower the data rate that can be supported.

The data rates available are also dependent on the national standards and are typically multiples of the ISDN basic rate. This gives a slightly different maximum value in the US and different intermediate values where the rate is based around multiples of 1.536 Mbits per second. This gives supported rates of 1.536, 3.072, 4608, and 6.144 Mbits per second.

These rates allow an ADSL link to be added to an existing ISDN network, where primary rate data can be transferred down some of the channels within the ADSL.

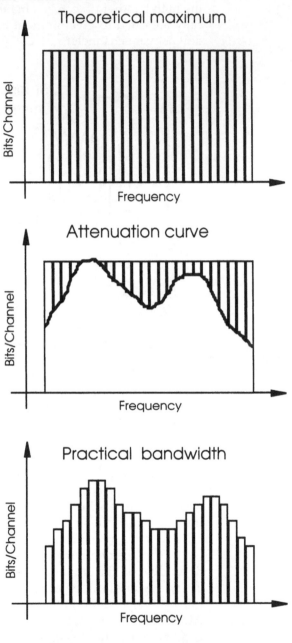

Effect of attenuation

Data rate	Cable length
2.048 Mbit/s	4.8 km (0.5mm Cu)
4.096 Mbit/s	4.0 km (0.5mm Cu)
6.144 Mbit/s	3.6 km (0.5mm Cu)
6.144 Mbit/s	2.7 km (0.4mm Cu)

ADSL data rates and cable lengths

Using ADSL

The technology that ADSL uses is not the whole story: it allows high bandwidth data to be delivered from the network using the existing analogue POTS cabling, whilst preserving the existing POTS service and equipment. In supplying this level of data bandwidth, it can provide support for the transmission of MPEG encoded video and, as a result, allows multimedia applications, such as videos on demand and interactive television, to be delivered to the consumer without having to replace the existing network.

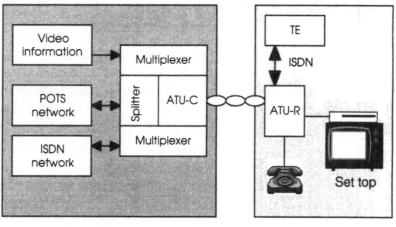

An ADSL implementation

This can provide ISDN compatible services, POTS telephone calls and video transmission simultaneously down the cable. However, the question arises of how this bandwidth is going to be made available within the network to support this level of data communications. Whilst ISDN does support high bandwidths with its primary rate specifications, even these are not high enough to support the demand forecast when video on demand and set top television becomes an everyday reality.

ADSL is part of the solution, in that it provides a network to subscriber delivery system. It does not provide the higher protocol levels necessary to ensure that data and bandwidth to support the applications are available.

If a new standard is being developed to support high speed data transfer, why not use the same standard as the next generation of local area networks for computers? If this is done, it would provide the opportunity for computers, multimedia devices and telecommunications devices to use the same network, irrespective of whether it is a public telephone network or a local area network in an office. This new standard is ATM. Before explaining how ATM works in Chapter 11, the next chapter will discuss the standards that are used by the computer world to achieve its networks such as Ethernet, Token Ring and FDDI.

Alternative DSL technologies

The idea of using ever advancing digital signal processing technology to increase the data bandwidth of twisted pair cabling has not stopped with ADSL, although this technology is likely to be the first to be deployed in trials and commercially.

High bit rate digital subscriber line (HDSL)

HDSL offers full-duplex E1/T1 data rates access over two twisted pair copper-wire pairs. The specifications were created by Bellcore and use 2B1Q modulation techniques and do not require repeaters to achieve the data rates over the specified distances. The technology is frequently used to replace repeated T1 service over distances as long as 3500 metres. Unlike ADSL, HDSL provides users with 1.5 to 6.1Mbps on each stream and thus creates a two way symmetric communication full duplex link. This lends itself as a candidate for delivering high speed data links into offices and the home.

Symmetric digital subscriber line (SDSL)

Also known as single-line digital subscriber line, SDSL offers E1/T1 transmission speeds in upstream and downstream directions over a single copper-wire pair. SDSL is full-duplex, so it provides speeds of up to 1.5 Mbps both upstream and downstream.

The technology may be the preferred method for doing sophisticated real-time functions, such as conducting audio, data, and video communications, or remotely connecting to a corporate LAN.

Very high bit rate digital subscriber line (VDSL)

An impending upgrade to ADSL, VDSL combines ADSL technology with ATM to give users speedy communications and network access over a twisted-pair copper wire at speeds of up to 60 Mbps downstream and 2.3 Mbps upstream over distances of up to 300 metres. This lends itself to the delivery of video and is likely to be the technology used for video on demand. The high speed data rates do not work over large distances and in practice, its deployment requires fibre optic links to provide the high speed infrastructure.

ADSL lite or G-Lite

This has been developed by Compaq and Intel and is essentially an ADSL or an SDSL link that goes straight into a PC and therefore does not necessarily need a splitter to generate the POTS connections. Any POTS facilities would be provided by the PC itself. This idea is gaining ground but its acceptance may rest on other issues such as telecommunication regulations concerning exactly where equipment can be connected to the public network. In the USA, it is clear that this is less of an issue and the idea of a digital connection via the existing copper wire infrastructure that offers 1.5 Mbps data rates is attractive. From a network perspective, the ability to deploy the technology without having to have a technician install a splitter is a great saving on cost. However, it does rely on the public being able to connect direct onto the network. In the past, this has been a matter of great concern with the telecommunications companies who have maintained strict control over any equipment that connects to their network. However, many millions of modems are connected onto the public networks everyday and there is a strong argument that using a splitterless ADSL or SDSL connection is no different. It will be interesting to see how this technology is deployed and whether, as many predict, it will be an initial success in the US first.

10   LANs

This chapter describes the various standards and technologies used to implement local area networks or LANs.

LANs are becoming important because most computers now used in business applications are connected to a LAN and therefore, for many users, it is the delivery medium of choice for multimedia data. Unfortunately many LAN technologies, and Ethernet in particular, were developed for the transfer of simple data and not video and/or audio. As a result, there is an immediate problem of using an existing system which is not suited to the traffic it is now being asked to support.

The main problem areas are the increased volume of data that needs to be delivered for multimedia and the requirements for real-time and predictable data transfers. For many existing systems, the bandwidth offered is only just sufficient for data traffic, let alone MPEG movies. As a result, multimedia is often not implemented on LANs that could potentially support the technology. The real-time aspects are very important: who wants to watch a movie and hear the speech a second later because someone else on the network was copying a file?

To understand these issues, it is necessary to understand how LANs are implemented and how they operate. The first standard covered is the most widely used —Ethernet.

Ethernet

Ethernet is probably the most widely used network system for PCs. It was developed by Xerox (who also pioneered the graphical user interface found in the Apple Macintosh and with Windows) and Digital Equipment Corporation. It has gone through several phases since it first appeared, especially in the cable types supported and the controlling software used to move data across the network.

It is important to remember that the term Ethernet strictly refers to the method of transferring data and not to the complete OSI stack mechanism. However, this more formal definition has become worn at the edges and the term *Ethernet* is often used to describe the whole network, including the networking application itself. It should be remembered that Novell, AppleTalk, Lantastic and Windows for Workgroups can all run over an Ethernet network – but they are different systems and are not compatible. To use a car analogy, it is like saying you drive a Ford – the statement does not give any indication as to whether the vehicle is a car, a van or a truck.

Ethernet is considerably faster than serial protocols and transfers data at rates up to 10 Mbits per second (Mbps). This may seem extremely fast, when compared with the 115 kbits per second (kbps) that can be achieved with serial lines, but the actual throughput is considerably less than the 100 fold improvement that the theoretical maximum would indicate. There are two reasons for this: the first is the overhead of the networking software in preparing and receiving the packets of information that are sent across the network and the second is the protocol used to arbitrate between nodes simultaneously trying to broadcast information.

CSMA and its implications

CSMA/CD – Carrier Sense Multiple Access with Collision Detect – refers to a mechanism used by the Ethernet protocol. It was derived from the Ethernet specification developed by Xerox in 1975 at its Palo Alto Research Centre (PARC) and designed to stop message corruption when two or more nodes transmit data simultaneously. This is necessary to prevent data traffic from being corrupted by too many nodes transmitting simultaneously.

The main characteristic of CSMA/CD is that any station can send a message at any time, irrespective of the line status. If a station wishes to send a message, it listens to see if the line is busy. If it is not, it sends the message and listens to check that no one else is transmitting and therefore corrupting the data. If someone is simultaneously transmitting, a collision is detected and the message jammed to indicate to the recipient that this has occurred. The sender then backs off the line, waits a random time and attempts to re-transmit the message.

If no collision is detected, the sender keeps transmitting until the complete message is sent. If the line is busy, the sender waits for a random time and tries again. So, if a node only transmits when the line is quiet, why have the collision detection in the first place? The answer is transmission delays within the network cabling itself – because it takes a finite time for a data packet to travel from one node to another, it is possible for a node not to hear a message, simply because it has not reached it yet. The node therefore wrongly assumes that the line is quiet, only to discover the earlier message arriving and corrupting its own transmission. This is why the CSMA/CD mechanism is needed.

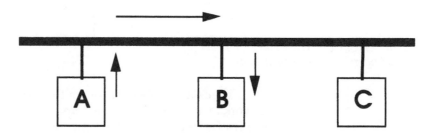

A transmits to B. C does not hear transmission

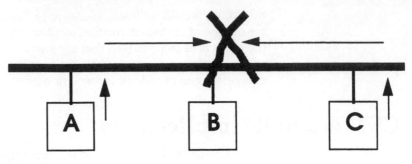

C starts to transmit and causes a
collision. A and B both stop and
wait.

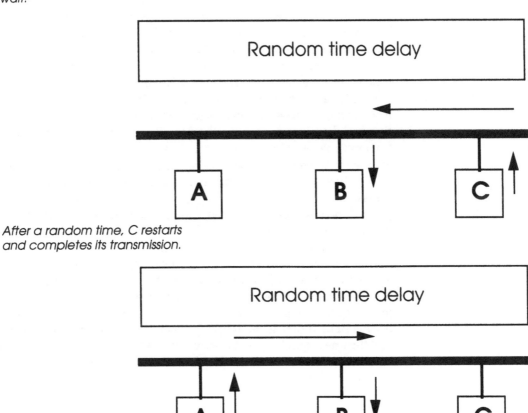

After a random time, C restarts
and completes its transmission.

After a random time, A restarts
and completes its transmission.

The delay involved in transmitting a packet from one end of the
network to the other effectively imposes a limitation on the physical
length of a network. If the network is too long, the CSMA/CD
mechanism ceases to work. This maximum distance depends on the
nature of the cabling used for the physical layer — i.e. twisted pair,
coaxial cable etc.

CSMA / CD is also indeterminate, i.e. the time taken to send a message can vary and, in some cases, it may even not be received because of the large number of collisions caused when many nodes try to use the network. Every time an attempt is made, the message can be aborted through a collision until either the node times out or the message's relevance is lost. The threshold for such performance degradation is dependent on the number of nodes, message length and the number of attempted accesses per unit time. Once this threshold has been reached, overall system performance and through-put degrades dramatically. A maximum of about 60 nodes is recommended for many networks, although heavy utilisation can start affecting the system with as few as ten nodes. These characteristics effectively rule out Ethernet for real-time process control applications, as with such a system it is impossible to guarantee when or even if a message is actually received! However, for data interchange, this can be tolerated. The key to implementing an effective Ethernet network is to control the data traffic through very careful design.

Ethernet standards

There are three main standards connected with the term Ethernet: the Ethernet versions 1 and 2 and the IEEE 802.3 standard. Unfortunately, these standards differ slightly and are therefore not really compatible with each other.

Ethernet 1 was the original design based on work performed by Xerox at PARC in 1975. Ethernet 2 (the current Ethernet specification) first became available in 1982. However, the IEEE published a set of LAN standards in 1985 covering the IBM Token Ring, the MAP (Manufacturers' Automotive Protocol) interface and an Ethernet type LAN called 802.3. The main differences between these two Ethernet types are a slight difference in frame formats and in how the different communication levels are implemented. With true Ethernet, the lower two levels of the OSI stack are treated as a single Data Link layer and in addition a special Ethernet Configuration Test Protocol (ECTP) exists. With IEEE 802.3, the Data Link layer is replaced with separate MAC and LLC layers as per the OSI model. ECTP is also not supported.

Although these specifications are very similar, they are incompatible and many configuration problems are caused by some nodes using true Ethernet while others are using IEEE 802.3. Many adapter cards support both and the standard is selected either by software or by hardware jumpers. As the differences are at the lower levels, network software (such as Novell Netware) does not have any preference over which standard is used and so moving a Novell Ethernet node from one network to another can fail because one network was using Ethernet and the other IEEE 802.3.

Cabling options

A number of cabling options are available for Ethernet based networks, which offer different installation costs and characteristics. Cabling is either connected directly to each interface board or through some kind of a transceiver. The transceiver approach has several advantages in that it allows the same interface card to connect to different media simply by changing the transceiver. However, it is common for interface boards to include a combination of BNC (thin coax) and UTP transceivers on board, thus providing direct links. The one exception to this is with thick Ethernet.

The transceiver interface has been standardised and uses a 15 pin D-type connector and is known as the AUI (Attachment Unit Interface) port. Unfortunately there are some slight differences between the IEEE 802.3 and Ethernet 1 and 2 standards and it is important to use compatible transceivers and boards. To make matters worse, Apple have a different standard for their own Ethernet boards called AAUI (Apple Attachment Unit Interface), which uses a different connector.

10Base-5 – 'thick Ethernet'

Known as the infamous 'yellow peril' due to its colour and lack of flexibility, 10Base-5 was, until relatively recently, the only way of cabling an Ethernet network and this cost often inhibited its use. It can be used up to distances of 500 metres and its use today is often restricted to providing a central backbone for a large network.

As stated previously, nodes are connected to the cable through the use of special transceivers or taps as they are known. Taps is a good description of these units as they physically cut their way through the sheaths to make contact, in much the same way as a plumber or electrician will tap into existing pipes or circuits. These units are expensive and as a result, thick coax cabling is usually reserved to forming a backbone or where its superior performance is needed.

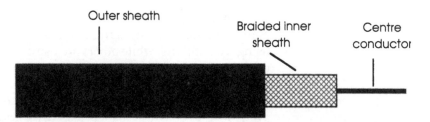

Coaxial cable

There are two types in common use: the first uses screw connections to effectively break the cable to insert the node and the second type clamps around the cable and inserts a spike through the sheathing. This type is affectionately known as a 'bee sting' from the shape of the spike. In both cases, a second cable – known as a drop cable – is used to link the transceivers with the interface card. In addition, thick coax networks must be correctly terminated to prevent the transmitted signal from being reflected and causing interference.

Thick Ethernet has rapidly become the least favoured cabling option because of the higher cost of adding nodes and the relatively inflexible nature of the cabling. It does provide good long distance links and potentially had a place in the backbone of a network. It is still however an unpopular option and should not be considered for new networks.

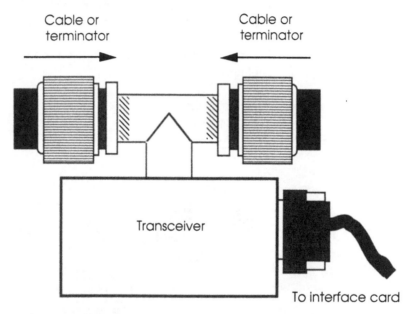

Screw fitting coax transceivers

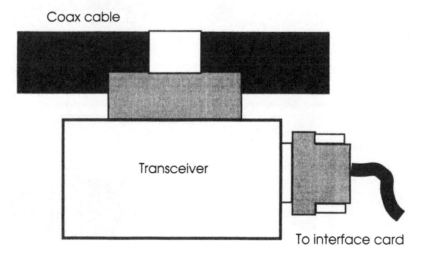

Clamp coax transceivers

10Base-2 – 'Thin Ethernet'

Thin Ethernet is often referred to a 'cheapernet' as it is significantly cheaper, more flexible and easier to install than thick coax. It
works over a smaller distance — up to about 200 metres — but it does
not need any special transceiver or tap using a standard BNC tee
connector instead. Although not quite as good as thick coax, its
performance is better than UTP and it does not need special hubs. For
many small installations, it is on par in terms of cost with a similar
UTP based network, and is probably easier to install.

Thin coaxial systems can easily be terminated by special BNC
connectors which have the termination network built into them. This
termination is necessary to ensure that the data transmitted down the
cable is not distorted by signal reflections and other similar problems.

For small installations, it is a very cheap way of installing
Ethernet because it does not require a central hub, unlike twisted pair
cabling.

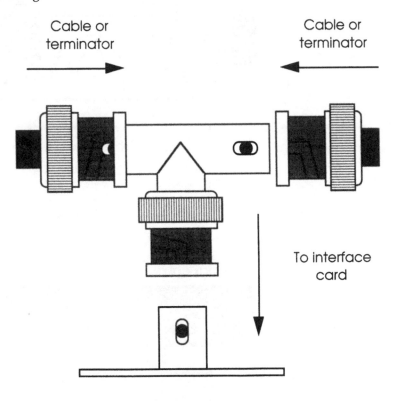

Cable or
terminator

Cable or
terminator

To interface
card

A BNC tee piece

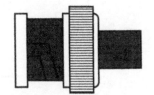

A BNC terminator

10Base-T – 'UTP'

This is the lowest cost cabling available for Ethernet and uses two twisted pairs to transmit the data traffic. It is similar to telephone cabling and its initial popularity came from the fact that existing telephone cabling could be used to implement an Ethernet link.

It is now a frequently used cabling for Ethernet and Token Ring networks although the performance in terms of cable length and its low tolerance to external interference such as fluorescent lighting can impose limits on its use. Although the cabling is low cost – especially compared with thick and thin coax – it does need a special concentrator or hub which does dramatically add to the cost.

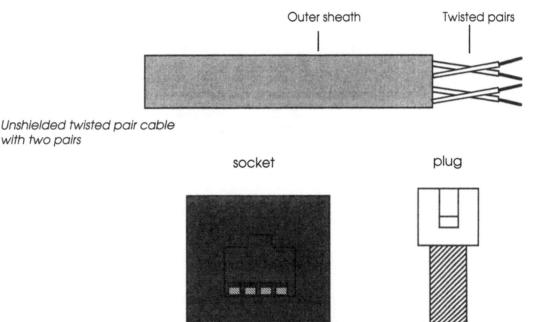

*Unshielded twisted pair cable
with two pairs*

A RJ45 type socket and plug

The American RJ45 style of plug and socket is often used with this cabling.

Fibre optic cable

Once regarded as 'high tech', fibre optic cabling is fast becoming popular, especially in areas where security is paramount. The data is sent down the cable using a beam of light and does not generate any electrical interference from the cable. In addition, unlike electrical cable, it is virtually impossible to break into the circuit without destroying its integrity. It is more expensive and harder to install, although its cost is rapidly decreasing and the installation techniques improving. Unfortunately, at present, there are no standards for this cabling type.

Termination and cable breaks

With Ethernet's relatively high data rates of 10 Mbps, it is important that the cabling is correctly terminated to prevent reflections of the data being generated and causing signal degradation. This is normally done by terminating each loose end of the network with a 50Ω or 75Ω resistor. Such terminators can be bought as plugs for thin and thick Ethernet and are built into the hub for twisted pair. However, if the network is broken in half, either through a break or accidental decoupling then there are two potential problems to resolve.

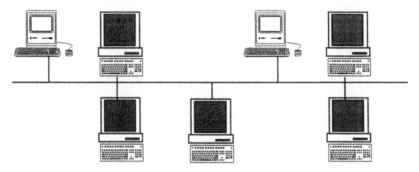

The original network

Part A   Part B

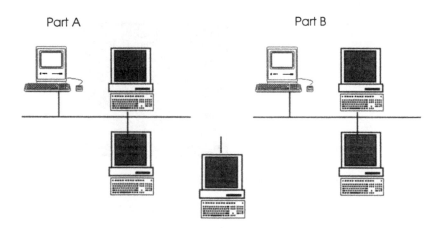

The network after disconnection

The first is that the network has now been split into two smaller independent networks. If a node was using a server on the other half of the network, that link would be immediately lost. If the server was being read from, apart from the inconvenience of losing the link, nothing else is likely to be lost. If the server was being written to, there is a possibility that the file will be corrupted. The second issue concerns the lack of termination at the two ends formed by the break. Without termination, there is a possibility that sufficient reflections will be generated to interfere with the data traffic and effectively stop any communication on the smaller network.

Either way, the simple break in the network cable can cause immense problems. However, there are several solutions. With a 10Base-T network, each node is independently connected to a hub and can be disconnected (either physically or electrically) without affecting the other nodes — unless the disconnected node is acting as a server!

The second solution to the break problem is to use special sockets which perform a 'make before break' operation. It short outs the connection so that the network is not broken when the node is removed. This does not prevent breaks caused by cutting through the cable or other such activities, but it does prevent disconnection from causing a problem.

The third solution is to use self-terminating connectors. If a break occurs, the two new ends are automatically terminated and each new half will function correctly. This does not solve the potential problem of the lost server but it does allow the remaining nodes to continue working.

The disadvantage of self-terminating and make before break connectors is their cost, which can be several times higher than the normal connectors. For this reason (and especially on large installations, where it is often impractical to check every node before removing or disconnecting a device), 10Base-T is often a more cost effective solution. If the network is fairly small or the nodes are fixed desktop PCs, there may be no need for the extra cost of these specialised connectors. It is simply a case of deciding where priorities are and how much a network break would cost in lost work, business, revenue and inconvenience.

Ethernet packets

Ethernet really only describes the mechanism used to transmit packets of data. Having an Ethernet interface card installed in a PC and connected does not mean that you can communicate. To use an analogy, all that has happened is that you have worked out how to dial a telephone number in France. Unless you can speak and understand French, it is highly unlikely that any meaningful communication will take place! So how does Ethernet take data and transmit it ?

Data is encapsulated into a specific packet or frame which has 6 sections or fields. Two types of this packet are used: Phase 1 or Ethernet and Phase 2 or IEEE 802.3. It is important that a network uses the right type. Phase 1 and Phase 2 packets are slightly different and this can cause problems.

The preamble fulfils two functions: it acts as a signal to other nodes that another node has started transmitting and that other senders should wait and it identifies the start of a packet. The destination address identifies the physical address of the network node that the packet is intended for. This address is held in four pairs of hexadecimal numbers. Each Ethernet card or interface has a unique address. Some networks require the installer to manually enter this address while others automatically interrogate the board for it. All nodes read every packet and this field is used to determine whether the packet is addressed to them or not.

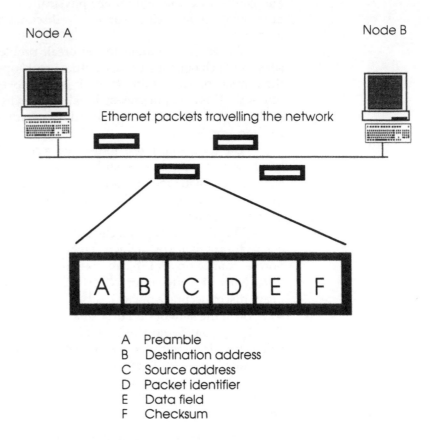

Node A

Node B

Ethernet packets travelling the network

| A | B | C | D | E | F |

A   Preamble
B   Destination address
C   Source address
D   Packet identifier
E   Data field
F   Checksum

Ethernet packets

The next field contains the address of the sender (or source). This allows the receiving node to identify who sent the packet and therefore who to respond to or ask for more information. The fourth field has different uses, depending on whether the packet is phase 1 or phase 2 compatible. With a phase 1 packet, the field is used as a type field and allows different networking software to tag a packet so it can be easily identified. This is advantageous where several different networking systems work over the same cable. With a phase 2 packet, this field contains the length of the data field. The data field contains the data being sent. If the data is less than 46 bytes, it is padded out. If it is more than 1,500 bytes, it is broken into smaller 'chunks' and sent using multiple packets. The final field is a checksum, used to ensure that none of the received data has been corrupted. The receiving node calculates the checksum from the incoming data and compares it with the transmitted value. Any discrepancy and the packet is deemed as corrupt and discarded.

It is important to remember that there is nothing to stop other packet formats from being incorporated into the data field of the

Ethernet packet. Most networking software does exactly this to provide network independence. For example, a Novell Netware network can work over Ethernet or Token Ring simply by encapsulating the data with its own protocol information in an Ethernet or Token Ring packet. When the packet is received, the Novell data is removed from the Ethernet or Token Ring packet and presented to the software. In this way, the Novell software does not have to be aware of how the data has been transmitted. This same technique is used in Apple's AppleTalk networks. AppleTalk packets are inserted into Ethernet packets when the networks use Ethernet.

Ethernet performance and its limitations

Ethernet has established itself as the de facto standard for PC networks. Although the individual packets of data travel at 10 Mbps, this is not a sustained rate and the overall throughput is much lower, due to collisions and the overhead of putting the data into the packets. To give some idea of the differences, here are some measurements of the time it took to transfer two files from one hard disk to another using Ethernet, LocalTalk and locally on an Apple Macintosh computer connected to a small network. (LocalTalk runs at 230 kbps.)

	Ethernet	LocalTalk	Disk
Transfer 540 kbytes	19.2	53.04	4
Transfer 4 kbytes	2.2	2.3	2.04

N.B.   *All times in seconds*

Network timings

The first thing to notice is that with the smaller file, there is almost no difference in the times. This is because very few packets are needed to transfer the data and the actual transfer time is small compared with the time needed to locate and create the file and its directory entry. However, with the larger file, LocalTalk is almost three times slower than Ethernet, although Ethernet offers over 40 times the throughput . . .

There are several reasons for this apparent discrepancy. First, the results measure the time to perform a *complete* file transfer and not the maximum throughput of the network. In this case, the system at the receiving end could not accept the data fast enough to keep up, so the Ethernet network had to wait, thus reducing its overall throughput. Secondly, Ethernet packets are not transmitted back to back and the separation time further reduces the overall throughput.

However, when networks are loaded, Ethernet performance is more consistent. Instead of waiting, the available Ethernet bandwidth can be used to transmit data between other nodes without affecting the original transmission. This is not the case with LocalTalk, which saturates extremely quickly because a node will accept the data from the network as fast as it can be sent.

Whilst Ethernet does not deliver the full 40 times performance improvement, it can transfer data faster than many hard disks and, more importantly, does not saturate as quickly.

Although in small networks, Ethernet will transfer data fairly quickly and consistently with a variable loading, as the network gets bigger and the number of data collisions increases, its performance decreases dramatically. The point of onset of this reduction depends on the nature of the data being transferred. The larger the files, the earlier saturation will occur. In such cases, it may be better to split the network into several smaller ones using a bridge to filter the traffic, reducing the actual load on each segment of the LAN.

Ethernet and multimedia

For multimedia applications, it is important that bandwidth is available and guaranteed so that the data stream can be maintained. With Ethernet, this is not the case, and depending on the loading (and hence the number of collisions on the network), the time taken for a data packet to be successfully transmitted can vary. For an audio-video display reliant on data from a hard disk on the network, this can cause problems when the data is needed and not available. To solve this, buffering is used to provide a reservoir in case packets are delayed. The buffer can hold many packets — the number being dependent on the worst case delay that the system must tolerate and the resulting delay in starting the playback — and only start the playback once the buffer is full. If packets are regularly received, the buffer stays full until the end packet is received. If a packet is not received when expected, all that happens is that the buffer is emptied by one packet. Various handshaking protocols can be added to this basic mechanism to ensure that the data supply to the playback system is maintained.

This method appears fine, but the delay in filling the buffer can be inconvenient if it is too long (users tend to expect something to happen immediately and not two or three seconds later!) and, more importantly, it is a major issue with interactive or synchronised multimedia, such as video conferencing. Imagine trying to hold a simple telephone conversation with a two second delay between the two parties. In practice, the acceptable delay for good video conferencing is about 150-200 ms.

To prevent this problem, bandwidth must be guaranteed or some method for sending priority data needs to be used to ensure that the buffers can be kept to as small a size as possible. As a result, many developments in the LAN arena are now focusing on supporting these requirements as well as providing higher bandwidths.

Token Ring

The token passing ring, IEEE 802.5, is mainly used within IBM networks. It is often referred to as Token Ring and is similar to the

MAP protocol 802.4, in that a token is passed between nodes on the ring to control access, but differs in that the ring is physical rather than logical. Each node receives data and passes it on. Possession of the token is a prerequisite for transmitting data. This provides a collision free network, allowing high transfer rates. The main problem comes from maintenance of the ring's integrity in case of node failure, disconnection or introduction. Relays are often used to provide ring bypasses, allowing the removal of nodes.

If a node wishes to transmit data, it captures the circulating free token, sets the 'token busy' bit and transmits a data frame. This frame circulates around the ring and eventually returns to its originator. The originator strips off the token busy bit, freeing the token for another node to use.

Basic principles

Shielded twisted pair cable medium is often used for Token Ring. This uses a hub to give a physical star configuration, which may seem a little strange at first because it does not look like a ring at all. However, each link to the hub is made using two cables, in and out connections to the node, so the physical cabling still maintains the ring topology.

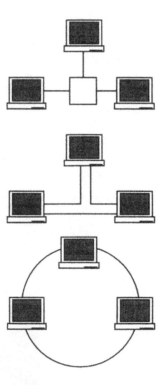

Using a star topology as a ring

The central hub is known as a multi-station access unit (MSAU) or media access unit (MAU). This device is responsible for passing transmitted frames from one node to the next. Each node has a special Token Ring network adapter card which receives a data frame from its upstream neighbour and re-transmits it to its downstream neighbour. The frame is passed around the ring by each station receiving and re-transmitting it. The MAU is responsible for routing the incoming frames out to the next station and for preserving the data transmission integrity. Frames are circulated around the network at either 4 or 16 Mbps, depending on the hardware used.

The frame may either be data being sent to a destination on the network or just a token. If a node wants to send some data, the network adapter waits for the token to arrive, appends the data and destination address to it and re-transmits the new frame. This frame circulates around the network until it arrives at its destination, where the data is accepted and an acknowledgement added. The acknowledged frame is transmitted around the ring until it reaches its originator, which converts the acknowledged frame back into a token, ready for another node to use.

Using this token method solves the problem of multiple nodes accessing the network simultaneously in a more elegant way than the CSMA/CD process used by Ethernet. Time is not wasted in aborting frames, waiting and re-transmitting.

Connectors and cables

The original IBM cabling standards specified two types of Token Ring cables made from four wire shielded twisted pair cables to act as an adapter cable from the interface card to the MAU and a ·second patch cable for installations where a workstation could not be located 8' from the MAU. Since then, several alternatives using RJ-11 or RJ-45 modular telephone connectors and unshielded cable have become available.

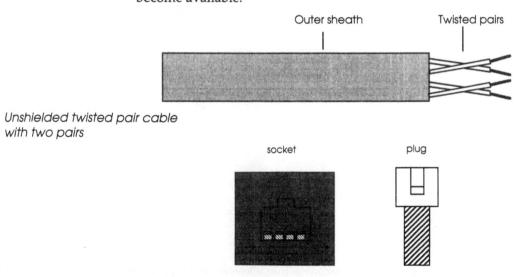

*Unshielded twisted pair cable
with two pairs*

A RJ45 type socket and plug

It is important to remember that with Token Ring (unlike Ethernet) these connection cables carry a constant current and are polarity sensitive. The constant current allows the MAU to monitor the ring and jumper across a disconnected node connection to preserve the ring's integrity. If the ring is broken, no one can transmit data and all communication within the network will fail. While the connectors used are keyed to prevent incorrect connection, care must be taken when wiring a cable. Reversing the connections on a twisted pair could damage the adapter card and the MAU. (This is becoming a dangerous trend with other hubs that support other standards also using the same connector with different electrical connections. This can exist with Iso-Ethernet hibs for example.)

It is possible to daisy chain multiple MAUs to create a large network but this can decrease the network response even if the additional users do not access the network.

Token Ring performance and limitations

The performance of a Token Ring network depends on several factors. The first is the clock speed used to send data around the ring. Earlier implementations ran at 4 Mbps; newer networks normally operate at 16 Mbps.

The second factor is the latency of the ring i.e. the time taken for a frame to circulate around the network. To discover why this is the case, the mechanism used to circulate data around the ring must be understood.

Latency

With a Token Ring network, data is always circulating around the ring, even if it is just a token which no one wants to use — in which case the circulating traffic is just a token frame followed by a series of idle characters. These idle characters are necessary to maintain the synchronisation of the network. Characters have to be transmitted on a regular basis because the ring acts like a pipe, where the idle characters push the token around. If a node wants to transmit data, it waits for the token, converts it to a data frame and transmits it out, inserting idle characters to push it around. When the destination receives it, the acknowledgement is included and the data frame continues around until the sender receives it again, so each data frame performs a complete circuit of the ring.

The latency is the number of idle characters that are inserted before the sender receives the frame back. This is typically 50-100 characters for a 4 Mbps ring but can increase to over 400 characters with a 16 Mbps version. The latency is made up of the delay taken to pass the frame through each node as it circulates and is therefore dependent on the number of nodes and the length of the frame. If the frame is big enough, the start of the frame will be received before the frame end has been transmitted. In this case there is no need for any idle characters.

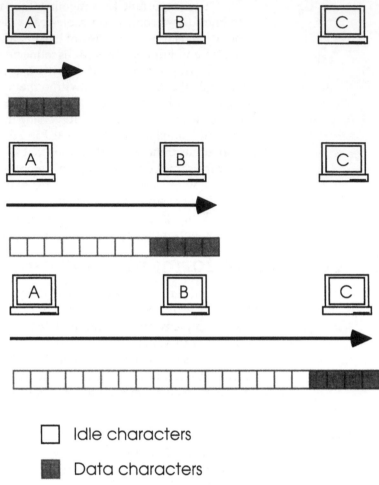

□ Idle characters

■ Data characters

*Sending idle characters to push
data along*

Early Token Release

Many messages and data frames are only a few bytes in length and latency can cause a big reduction in the effective use of available bandwidth. To solve this problem, a technique called Early Token Release (ETR) has been developed for use with 16 Mbps rings. The transmitting node sends a token after the data frame. As this new token circulates *behind* the data frame, any downstream nodes have an opportunity to use it and transmit data. The result is a network where multiple data frames and/or tokens circulate around the ring instead of just idle characters, increasing overall network performance.

Priorities and reservations

Each token has priority and reservation bits which allow a very sophisticated prioritisation system to be used. To use a token, a node must have a higher or equal priority to that specified in the token. If this is not the case, the token is passed downstream to the next node. The priority level of the token is set by the reservation bits in the current data frame which can, in turn, be set by any node that has a higher priority than the current setting. By doing this, the node effectively reserves the next access onto the ring for itself, unless a higher priority node needs it. To see how this mechanism works, consider the following Token Ring network as shown in the next few diagrams.

In the first diagram, a token is passed from workstation D to workstation A. The priority level of the token is set to zero and so A, with its higher priority, can take the token and convert it into a data frame for workstation D. The data frame is passed by workstations B and C as it goes around the ring until it finally arrives at workstation D. This is done by the insertion of idle characters that follow behind the data frame.

With the frame at workstation D, the data is read, an acknowledgement is attached to the frame and it is passed on. Workstation A receives the acknowledged frame and converts it back into a token before passing it on.

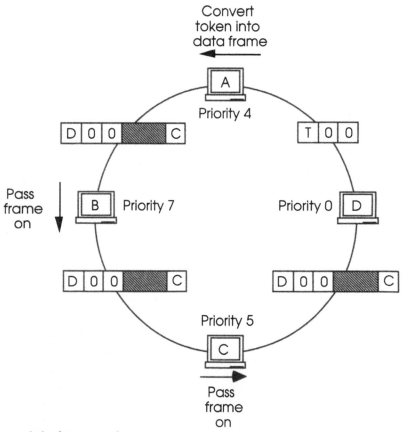

Converting a token into a data frame and
sending data to the destination

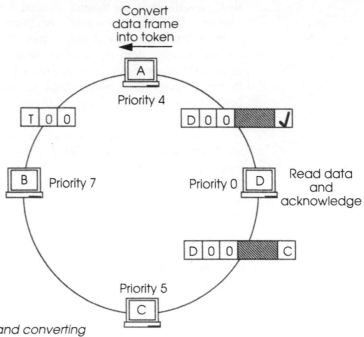

*Completing the transfer and converting
the frame back to a token*

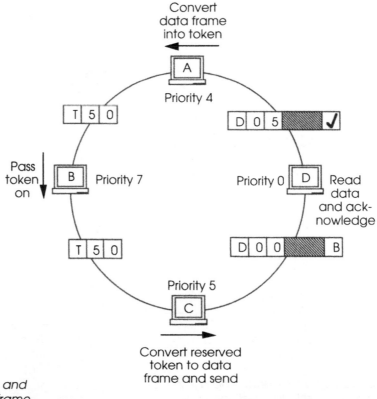

*Taking a reserved token and
converting it to a data frame*

The frame has no reservation bits set and the token is created with the lowest level of priority so that the first downstream workstation which needs access to the ring can use it. This is the basic process used to move data around the ring. The next example shows how the reservation bits come into play.

In this example, the token is received by workstation A as before, is converted into a data frame for workstation D and sent around the ring as before until it arrives at workstation C. C needs access to the ring as soon as possible and so it sets the frame's reservation bits with its priority level before passing the frame on to the next workstation D. Workstation D reads the incoming data frame, adds its acknowledgement and passes the frame to workstation A. As before, A converts the frame back into a token but, as the reservation bits were set to level 5 by workstation C, the priority level of the token is set to 5. The token now circulates. Workstation B could take the token because it has a higher priority level, even though it did not reserve the token. (If its priority level had been set to 2, it could not use the token even if it wanted to.) The token passes to workstation C, who reserved it on the last pass. C converts the token into a data frame and sends it out to continue its journey around the ring.

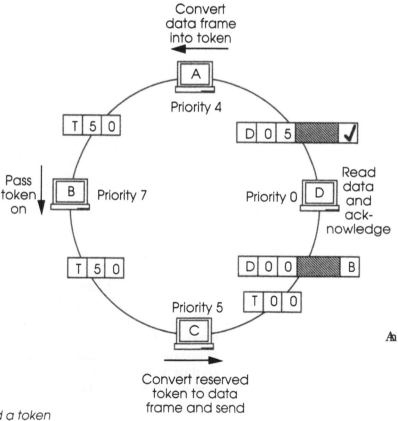

A ring using ETR to send a token after a data frame

With an ETR system, there would be a slight difference in the way the frames and tokens circulated. The examples above did not use ETR and so the next workstation to transmit would have had to wait until the data frame had completed its circuit and been converted back into a token. With ETR, workstation C would send out a new token after the data frame instead of idle characters and this would allow other nodes to transmit data more quickly, if they needed to.

Token Ring vs. Ethernet

With many network operating systems being capable of running on either Ethernet or Token Ring networks, the choice between the two has become less clear. The points in favour of Token Ring are its higher raw performance (16 Mbps as opposed to 10 Mbps for Ethernet) and the ability to prioritise loading through the priority levels and reservation bits. Ethernet, on the other hand, depends on random delays and waiting and its performance becomes less predictable as loading increases. However, Ethernet performance only degrades with the number of accesses and not necessarily with the number of nodes on the network. With a Token Ring network, the number of nodes directly affects performance.

In terms of cabling, Ethernet is relatively straightforward while Token Ring cable installation has been described as best left to professional installers because of its arcane properties. There is a view that it often requires a magic touch to make it work! This has become less of an issue as cabling systems have improved — but it is still something to consider. Perhaps the biggest differentiator is that of cost: network interface cards for Ethernet are substantially cheaper than those for Token Ring — although, again, these differences are reducing.

In summary, Token Ring is a good way to network but at a far higher cost than Ethernet, although within an IBM environment its higher costs may be justifiable.

Iso-Ethernet

Isochronous Ethernet (or Iso-Ethernet) is a practical solution to sending multimedia data across a LAN — particularly with video conferencing, which requires guaranteed bandwidth to function correctly. The main thrust behind it came from National Semiconductor, although it is now undergoing standardisation as IEEE 802.9a. In solving the problems of passing multimedia over a packet based LAN, it has several advantages:

* It is compatible with 10Base-T Ethernet and uses the same cable — thus removing any need for re-cabling.
* The additional isochronous channels are compatible with ISDN channels and therefore provide an easy interface to ISDN networks and ISDN based multimedia applications, such as video conferencing.

- The multimedia applications run independently of the Ethernet packets and therefore do not take valuable bandwidth away from the existing network.
- Only those workstations and hubs that use the additional channels need be upgraded within an existing Ethernet based network.

Isochronous Ethernet works by defining additional virtual channels which work in conjunction with the existing Ethernet channel. The diagram shows the virtual channels that are available in the bundle. The P channel supports the packet data that Ethernet sends and receives and provides 10 Mbps bandwidth, as normal. The C or circuit channel provides the equivalent of up to 96 ISDN B channels, with each B channel running at 64 kbps, giving a total of 6.144 Mbps. The 96 kbps D channel is used for signalling and providing similar facilities to those found in an ISDN D channel. The final addition is the M or maintenance channel that is used to transfer physical layer and control information.

The circuit channel, in conjunction with the D channel, can be arranged to imitate various ISDN configurations such as a BRI (basic rate interface) consisting of 2B channels and one D channel or a European primary rate of 32 B + 1 D channel.

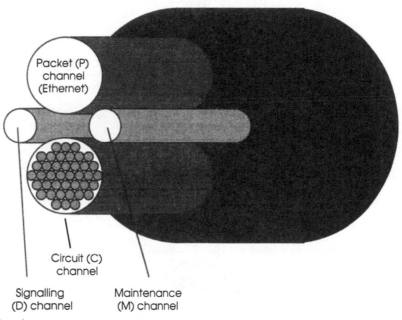

Iso Ethernet channel structure

The encoding used by the isochronous Ethernet standard is 4B:5B encoding, which is also used in FDDI networks.

The diagram shows an example Iso-Ethernet installation. The Iso-Ethernet hub is connected to four workstations with three of them (workstations A, C and D) connected using the Iso-Ethernet standard.

These workstations would have special Iso-Ethernet interface cards installed and the physical wiring would be the same as that normally used for a 10Base-T Ethernet network. The fourth workstation B is connected using just 10Base-T and therefore does not have access to the additional channels that the other workstation can use.

In addition, the hub can support an ISDN link which can connect to an ISDN terminal or handset, an ISDN based conferencing system or even directly to the public network to allow full access to the PSTN (Public Switched Telephone Network). These connections all use the same RJ-45 plug but with different wiring and most hubs provide a set of sockets for Iso-Ethernet, another set for 10Base-T and a third set for ISDN connections.

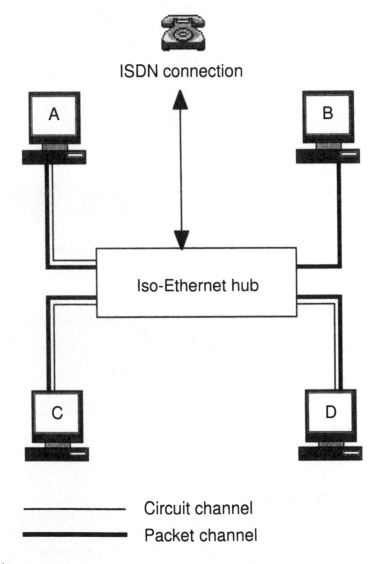

An example Iso-Ethernet installation

Iso-Ethernet is a clever solution to the problem of video over LANs. Whilst not requiring new cabling, it does require replacement interface cards for each workstation and replacement hubs within the network. The cost of these components is not insignificant, compared to standard 10Base-T versions. There is a further dilemma: for many information system managers, the last thing they want is load to be placed on the network — and while Iso-Ethernet puts video and other multimedia traffic on a separate channel, there is a fear that the other traffic on the Ethernet side of the network will dramatically increase. File sharing and access will increase as users find that they can work on files together during a video conference. With the promise of ATM and 'unlimited bandwidth on demand', many are following a sit and wait policy to see if this promise has any truth in it. With all this uncertainty in the industry, the adoption of Iso-Ethernet has stalled.

Switched Ethernet

Switched Ethernet is another attempt at extending the life of the Ethernet standard by providing more usable bandwidth. The idea is very simple and is based around the topology found with twisted pair networks, where a single node is connected to a hub. If the hub supported a higher bandwidth then it could allocate that bandwidth between the nodes so that each one had a full 10 Mbps bandwidth available, despite the bandwidth used by the other nodes.

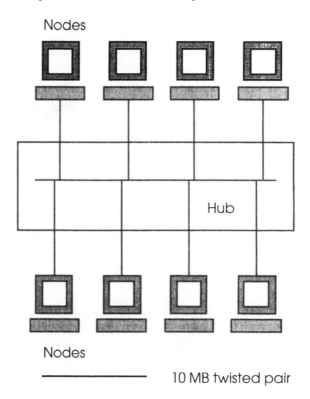

Nodes

Hub

Nodes

——————————— 10 MB twisted pair

Standard Ethernet hub

As far as the node is concerned, it can use the same interface card and sees a normal Ethernet link. The higher bandwidth is achieved within the hub and distributed to each switched Ethernet outlet and node via a bandwidth allocator. It is therefore possible to provide better performance simply by replacing an existing hub. This, and the fact that each node gets guaranteed bandwidth, has led to this technique being suggested as a way of addressing the limitations of Ethernet for multimedia. It is more expensive and, in large systems (especially where twisted pair cabling is not used), its cost advantage versus upgrading the system to a faster network is often unclear.

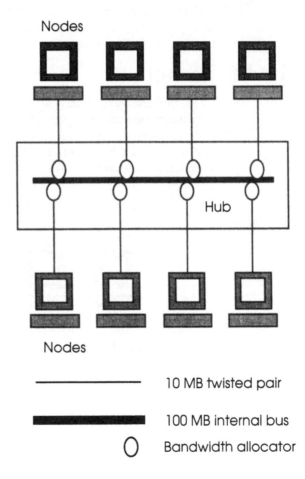

Switched Ethernet hub

FDDI

Fibre distributed data interface (FDDI) is a LAN technology for faster networks using fibre optic cable, although the specification has now been extended to encompass twisted pair cable. It supports large numbers of nodes with a bandwidth of 125 Mbaud giving transfer rates of 100 Mbps. This is done by using a special encoding technique called 4B:5B, where 5 level changes are used to encode 4 bits of data.

The protocol is very similar to Token Ring, in that a token is passed around a ring and a station must have the token before a data frame can be transmitted. However, FDDI does not use the idea of priority and reservation bits but instead uses a complex resource allocation algorithm to allocate resources and bandwidth.

Although expensive, especially when compared with Ethernet, its higher bandwidth is ideal for networks where graphics information is distributed or where there are a large number of users which are imposing heavy traffic loads.

The initial standard is in the process of being extended and is now known as FDDI I, with the extended standard called FDDI II.

FDDI basic principles

Like Token Ring, FDDI uses a circulating token to control access to the network. The token is transmitted as a frame that circulates around the ring when all nodes are idle. If a node wishes to send a data frame, it waits until a token frame circulates through. It takes the token by not passing the frame down stream to the next node and then starts to transmit data frames which circulate to their destination.

Other nodes wanting to transmit data must wait for the token frame to reappear. This is done by the node that holds the token, which regenerates the token frame and sends it downstream when it has finished transmitting its data frames. With the token regenerated and circulating, other nodes have an opportunity to grab the token and thus access the network and send data. Without this token, they can only listen and receive and pass on data.

Data is transmitted in frames of variable length. Each frame includes delimiters to mark its beginning and end, plus address information indicating source and destination stations.

The FDDI standard supports the concept of prioritisation by assigning a maximum time that a node can hold the token before releasing it. This time is based on its priority and, in this way, high priority messages can circulate quicker than low priority messages. This is a way of prioritising at the capacity instead of the frame level. With short messages that fit into a single frame, priority is effectively a round robin allocation, where each node in turn gets an opportunity to send data. The maximum times assigned to each priority do not apply because they are not invoked. The message is so short that the time-out never occurs and the message is sent in a single access. The worst case delay is the number of nodes multiplied by the time to transmit the frame. For large messages, a different picture appears.

With a large message that needs multiple accesses — i.e. the message is so big that the priority time-out is invoked and the token released with only part of the message sent — the number of accesses determines the speed and hence priority of the overall transfer. A node with the priority to send five frames per access is going to take twice as long as another node with the priority to send 10 frames per access if they both want to send data at the same time. This ignores the time to release and regain the token which has to be added.

With this capacity allocation system, the overall time to transmit data is prioritised but it will vary depending on the loading and compression on the network to use the node. The number of nodes, the amount of data traffic and so on are all part of the equation. As a result, communication rates between two nodes are not constant and this means that multimedia support is difficult.

FDDI II *extensions*

To overcome the problem of no support for constant data rate communication between nodes, extensions were added to the original FDDI I specification to create the FDDI II specification. This supports a circuit switched mode between two nodes and is upwardly compatible with the original FDDI I standards.

The new extension, often referred to as isochronous FDDI, imposes a 125 microsecond frame length to minimise delays. This is done by a single node, called the cycle master, sending a cycle code around the ring to impose the framing structure previously mentioned. This effectively defines a series of time slots that can be allocated for direct communication between two nodes. The allocation and connection are defined using the more normal FDDI I frames but, once established, the addresses within the frames are not used and the frames within a repeated time slot are directly allocated to the nodes by the prior negotiation.

When an FDDI II network starts up, it comes up as a compatible FDDI I mode where frame based communication using the possession of a token is supported. Once this has been established, the network switches into the FDDI II mode where the isochronous support is provided.

100 Mbps Ethernet

With Ethernet running out of bandwidth for many applications, and FDDI not becoming its natural successor, an initiative was started to expand Ethernet to provide compatibility but running at a higher bit rate and thus improve the available bandwidth. This resulted in the 100Base-T specification but unfortunately this has not found favour amongst the networking industry and an alternative called 100VG-AnyLAN has appeared as the favourite.

100Base-T

The 100Base-T or IEEE 802.3u, to give it its full IEEE reference, is Ethernet that supports bit rates of 100 and 200 Mbps. It uses the CSMA/CD model and this has lead to serious doubts within some parts of the industry over its capability to be scaled up to the new higher bit rates. The argument for the scaling is that the faster bit rates will mean shorter messages on the network and therefore less collisions. The case against it is that the assumption that message lengths will get smaller or stay the same is wrong and therefore, in practice, the CSMA/CD model will not scale very well to the rates that have been proposed. A further argument against 100Base-T is that the enhancements over Ethernet it offers, such as switching and full duplex operation, do not support the CSMA/CD model anyway. If this is the case, why support it?

It does use proven technologies: it uses CSMA/CD and the media access control layer from Ethernet and the physical layer is the same as used in FDDI.

Despite all this, it has other restrictions in that only one level of cascading is supported — which means that bridges are needed for additional levels. The maximum cable lengths it supports are in some instances 10 times less than supported under a 10Base-T system. This means that some 10Base-T networks cannot use 100Base-T as their upgrade path without having to re-lay the cable or add repeaters. It is probably this one factor that has caused 100Base-T's failure initially within the market place. However, it has subsequently been successful especially with the advent of adaptors that can support both the original Ethernet as well as the 100Base-T version. The cabling restrictions have also been worked around and for many new installations, 100Base-T is the first choice.

100VG-AnyLAN

With the initial doubts surrounding 100Base-T, several manufacturers, led by Hewlett Packard, developed an alternative 'fast Ethernet' called 100VG-AnyLAN. This has now been taken under the ISO wing as IEEE 802.12.

The base level transfer rate is 100 Mbps but rates of 400 to 2,000 Mbps are also specified. It is compatible with both Ethernet and Token Ring frames and therefore offers a smooth upgrade path for these technologies. The cabling is compatible with 10Base-T and can use the same cable. 100VG-AnyLAN uses approximately the same bandwidth as 10Base-T — 18 MHz against 15 MHz at a -3 dB low pass point respectively. The increase in data bandwidth is achieved by a change in the data encoding method used within the frequency bandwidth. One pair transmit Manchester encoding is used with 10Base-T to give 10 Mbps, while 100VG-AnyLAN uses four pair quartet coding using 5b/6b NRZ code to give 100 Mbps. The use of all the cables within the 100VG-AnyLAN specification does restrict operation to half duplex.

The physical topology used by 100VG-AnyLAN is that of a hub based star, although the logical topologies can be whatever is visualised by the higher levels of the network i.e. bus based for Ethernet, ring based for Token Ring, and so on.

Although the standard supports both Ethernet and Token Ring frames, it provides support for additional features that are particularly well suited to multimedia. It provides on-demand support (called demand priority access or DPA) where either Ethernet or Token Ring packets are transported directly from source to destination. In addition, packets can be assigned a priority level — and therefore are guaranteed to take precedence over lower levels of traffic and thus achieve better performance. This gives a big advantage for multimedia applications that need guaranteed bandwidth on demand. Such applications can be given higher priority over normal traffic and therefore guarantee a successful transfer where Ethernet with its CSMA / CD model cannot.

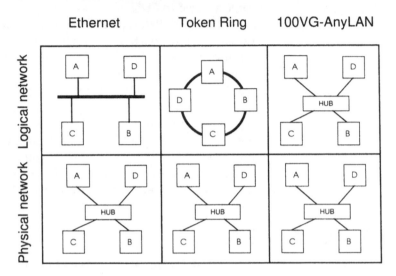

100VG-AnyLAN topologies

Operation

The operation is simply described in this section using the set of illustrations. In normal operation, the hub resolves any contention by using a round robin scheme, where each port is allowed to request to send a packet. This is normally signalled by the use of a low frequency signal.In the first overleaf diagram, workstations A and D send such a request to the hub. The hub uses the round robin scheme to decide on which one it should service first. In this case, workstation and port A get priority. The request from workstation D is effectively suspended, as shown by the greyed out arrow in the subsequent diagrams.

An acknowledgement is sent to workstation A which then sends the data packet. The framing for this can be either Ethernet or Token Ring and will include a destination for the data. In this case, the destination is workstation C. The packet goes through the hub and is sent out to workstation C. At the same time, depending on the type of transmission, the packet can also be sent via the uplink port. This port connects to other hubs to create cascades as shown.

After transmission, the packet is acknowledged and the next request from workstation D can be processed.

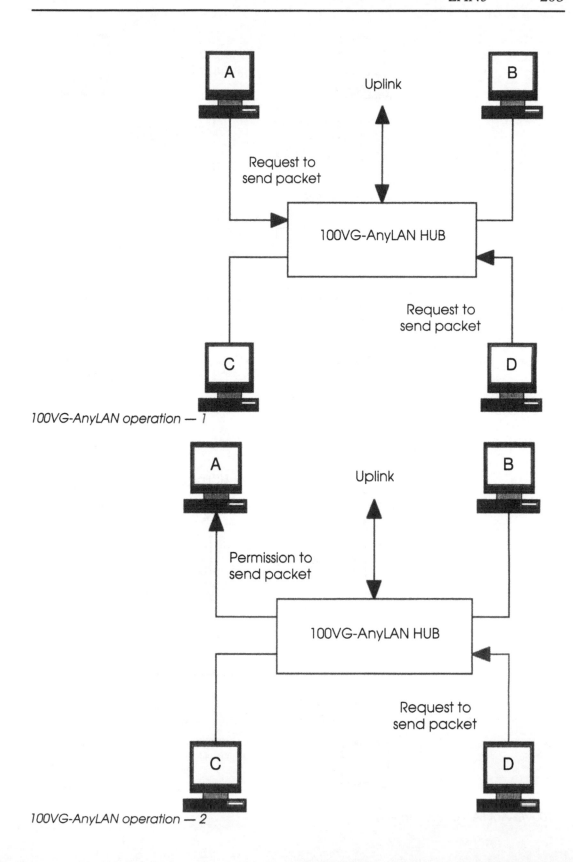

100VG-AnyLAN operation — 1

100VG-AnyLAN operation — 2

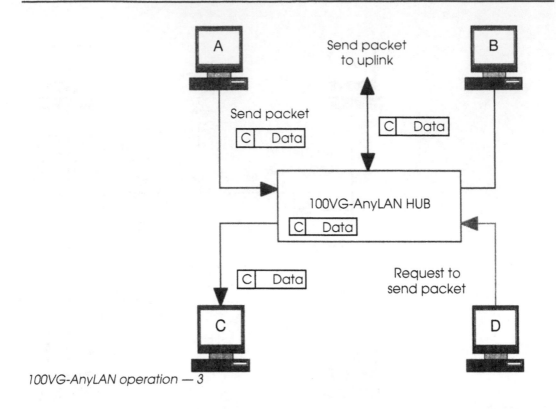

100VG-AnyLAN operation — 3

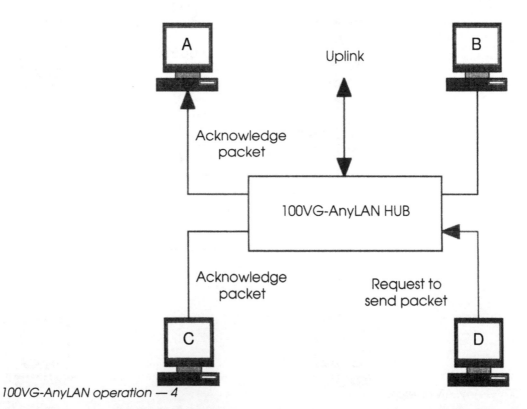

100VG-AnyLAN operation — 4

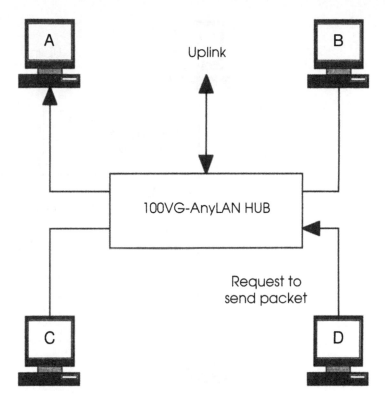

100VG-AnyLAN operation — 5

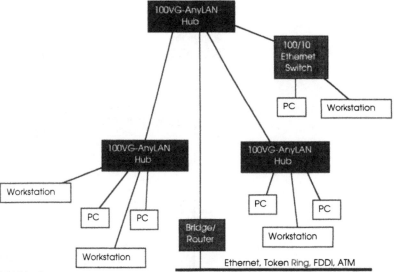

Cascaded 100VG-AnyLAN hubs

DPA operation

The DPA operation is similar except that the initial request is sent with a priority level which overrides the round robin scheme. If the request is the highest priority, it will take precedence.

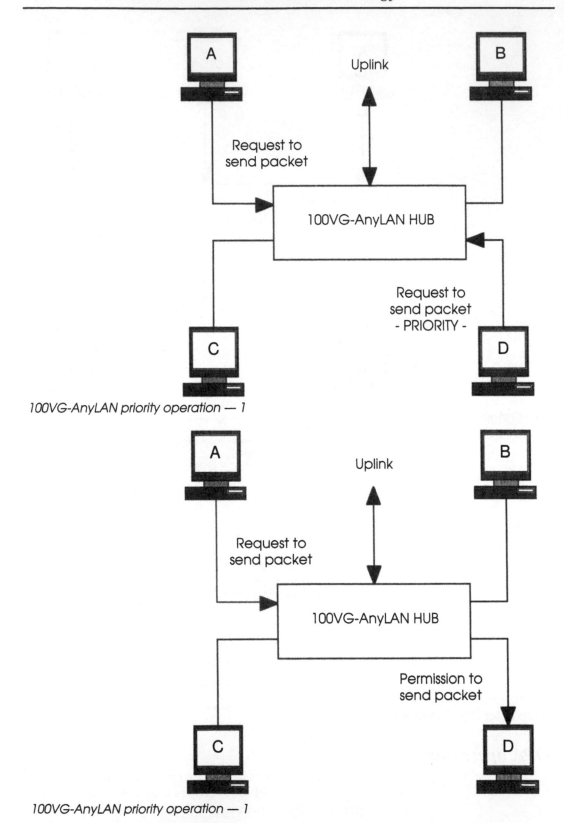

100VG-AnyLAN priority operation — 1

100VG-AnyLAN priority operation — 1

11   ATM

Asynchronous Transfer Mode (ATM) is a networking and transmission technology that was developed initially in 1982. It is based on an entirely new set of principles that enable such diverse traffic types as voice, data and video to efficiently share a common transmission path and switching points. It has received a lot of publicity recently and has been hailed as the panacea for all ills, providing unlimited bandwidth for all on demand — a little bit of an exaggeration.

It is an exciting technology that addresses many of the problems facing both LANs and WANs. It uses the same protocols, which makes linking the two networks together. It is a scaleable architecture that can run on many different speeds of physical media, such as twisted pair copper and fibre optic cables. It runs on top of existing delivery systems, such as ADSL, and supports the transmission of time critical data, such as video and audio.

The 'bandwidth on demand' phrase which has been associated with ATM is a little misleading. The ATM protocol will support this concept — but there is still a finite amount of bandwidth within the network and delivery system which obviously cannot be exceeded. What ATM does is to allow the ability to request and effectively pre-book bandwidth from the network so that a guaranteed delivery can be obtained. It is this type of service that is required for multimedia data transmission and it is this attribute that has caused so much interest. The basic principles behind ATM are as follows:

- ATM devices are connected directly to an ATM switch. This is typical in the telecommunications arena with a single wire per subscriber but is less common within the LAN arena. Coax based Ethernet links and Token Ring effectively share a cable between many nodes and thus share the bandwidth. The 10Base-T twisted pair implementation of Ethernet physically allocates a single wire per node but all nodes compete within the hub for the bandwidth that is available. The 10 Mbits per second (Mbps) bandwidth that each cable will support is still shared.
  Switched Ethernet is the closest LAN model to that used by ATM, in that each link from the node to the hub is used solely by the node and its bandwidth does not compete with other nodes feeding into the hub.
- All data in ATM is handled in containers, known as cells, that carry a fixed number of bytes. This principle has two important benefits. First, control over queuing delays is greatly improved, as the largest component of a queuing delay is caused by variation in data element size and ATM makes all data elements the same size. Second, equipment that processes fixed size data units is simpler than for variable-sized units.

- Each ATM cell has a header containing information that each ATM switching point uses in combination with previously stored routing information to determine where to send the ATM cell and thus transfer its information. This previously stored routing information is determined by the process of establishing a virtual connection or circuit. There is an entire set of machinery that maintains these virtual connections by manipulating the stored information at co-operating switching points. This machinery operates over the same network as the data, using its own virtual connections.

- Each switching point queues ATM cells for outbound transmission in a logical scheduling structure that allows control over such parameters as delay, delay variation and cell loss, based on the needs of the application, as expressed when the virtual connection is established. For example, a data file transfer is much more sensitive to cell loss than to delay, whilst a voice connection is more sensitive to delay and video to delay variation. This is referred to as Quality of Service (QOS).

- ATM cells are carried over transmission paths that can be of widely differing speeds. Currently, transmission speeds from 56 kbits per second (kbps) to over 9 Gbits per second (Gbps) are being standardised running on both copper and fibre optic cabling. Lower and higher speeds are foreseen.

The ATM structure

The structure for ATM is shown in the diagram. It effectively consists of three layers: the first layer, PHY, is concerned with the physical transmission of the data and depends on the medium being used. For each medium there is a different set of standards. The currently defined ITU standards and descriptions are shown in the table. It is misleading to think of ATM as a complete network as in reality it really only refers to the ATM and AAL layers. In the same way that Ethernet first appeared using thick coax cable and was developed to support alternative media such as thin coax and twisted pair, ATM defines a set of standards for existing and future physical interfaces. However, when most people refer to an ATM network, the physical connection is either mistakenly assumed to be 'ATM' or the physical layer or connection is simply forgotten.

The ATM layer is concerned with the maintenance of the cells and their routing through the switching network. ATM uses many small switches to route cells to their final destination and this information is stored in the VPI/VCI bits within the cell header.

The ATM adaption layer (AAL) provides standards for the segmentation and reassembly of data into the cells, and vice versa. This is also the layer that is concerned with the quality of service. The type of service is determined by asking for a particular AAL. There are five currently defined with some more under evaluation.

Convergence	CS	**AAL**
Segmentation and reassembly	SAR	
Generic flow control Cell VPI/VCI translation Cell multiplex and demultiplex		**ATM**
Cell rate decoupling HEC header sequence generation/verification Cell delineation Transmission frame adaptation Transmission frame generation/recovery	TC	**PHY**
Bit timing Physical medium	PM	

ATM structure

- I.113 Vocabulary of Terms for Broadband aspects of ISDN
- I.121 Broadband aspects of ISDN
- I.150 BISDN ATM Functional characteristics
- I.211 BISDN Service aspects
- I.311 BISDN General Networks aspects
- I.321 BISDN Protocol reference model and its application
- I.327 BISDN Network functional architecture
- I.361 BISDN ATM Layer specification
- I.363 BISDN ATM Adaptation Layer (AAL) specification
- I.364 Support of broadband connectionless data service
- I.371 Traffic and congestion control in BISDN
- I.413 BISDN User-Network Interface
- I.414 Overview of Recommendations on layer 1 for ISDN and BISDN accesses
- I.432 BISDN User-Network Interface-Physical Layer specification
- I.610 OAM Principles of BISDN Access

ITU standards

Above the AAL can sit further levels of software protocol, if needed and there are additional communication layers that provide other information, such as billing and bandwidth monitoring. The need for billing information is an obvious requirement and the standards are currently being defined. This means that an ATM link can currently only come from one network provider due to a lack of common billing data but this restriction will disappear in due course.

The data bandwidth requires monitoring to ensure that the user does not exceed the requested bandwidth and to allow accurate billing information to be sent both between the user and the network (UNI) and between various parts of the network (NNI). If the user exceeds his requested bandwidth, the network can either mark the cell as low priority, or drop the cell altogether. These may seem draconian responses but are needed to ensure that the network's other users do not suffer.

In addition, it is possible to impose other protocol structures so that ATM will emulate other connections and thus integrate easily into existing networks. An example of this is shown below.

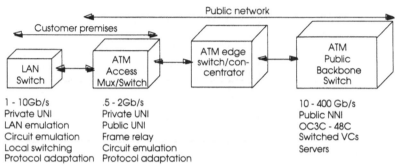

Physical networks for ATM

The ATM cell

The fundamental structure within an ATM network is the cell. All data is sent as cells that work their way through the network. Larger data structures are sliced into cell size packets and re-assembled; this is a function of the AAL. The cell structure is based on a 48 byte payload plus five bytes of header and routing information, making a total of 53 bytes.

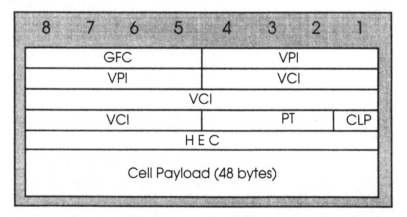

ATM cell description

The cell payload size is actually a compromise between a US requirement for 64 bytes and a European counter proposal for 32 bytes. The compromise figure that was reached was 48 bytes. This size is quite critical for two reasons. To support voice, the ideal is a small cell but if the cell is too small, the overhead of the header starts making large data transfers inefficient. A second reason is the potential problem with lines that are not fitted with echo cancellation. This is the case with the majority of lines in Europe and, to prevent problems, the cell size must result in a less than 20 ms transmission time on any one part of the line. Both the proposed US and European payload sizes met this requirement — but it did prevent larger payloads from being defined.

The header consists of the following fields:

GFC   The generic flow control bit only has local significance. Its value is not carried end to end. Studies are underway to investigate how this could be used for local user to the network interface communication.

VPI/VCI The virtual path/channel identifier or the protocol connection identifier (PCI) is used to control the switching and transfer of the cell. This is described in more detail in the next section.

PT   The payload type field uses 3 bits to indicate the contents of the payload. It is used to identify user data or connection associated layer management information.

CLP   The cell loss priority bit indicates if the cell is expendable. If set to one, the cell is low priority and, if the network is stressed or overloaded, the cell can be discarded.

HEC   The header error control field is an eight bit checksum for the first eight bits of the header. It can detect and correct single bit and some, but not all, multiple bit errors.

Routing ATM cells

Cells are routed through the network using a concept of virtual pipes and virtual connections. The ITU definitions for these concepts leave a little to be desired!

Virtual channel (VC)

A concept used to describe unidirectional transport of ATM cells associated by a common unique identifier value. This identifier is called the virtual channel identifier and is part of the cell header.

Virtual path (VP)

A concept used to describe unidirectional transport of cells that are associated by a common identifier value. This identifier is called the virtual path identifier and is also part of the cell header.

In practice, these concepts form a two level method of routing cells through the network. Virtual connections are carried by virtual paths and both can be switched to different paths and connections though VP and VC switches respectively.

The diagram shows the concepts between both types of switches. The VP switch on the left will redirect the contents of a virtual path to a different virtual path. The virtual connections it contains are unchanged. This is similar to switching an input cable to a different physical cable.

The VC switch has a set of virtual paths coming in and going out but these are not switched and are fixed. The virtual connections within the incoming virtual paths can be switched to different output paths by changing the virtual connection.

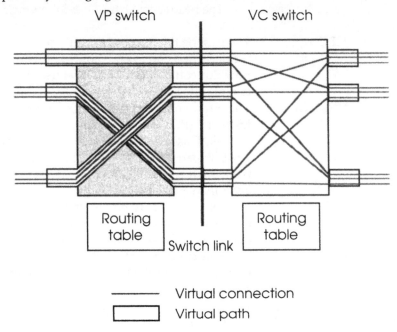

VP and VC switches

The combination of these two mechanisms provides two levels of switching. The clever part of this arrangement concerns how the switching information is actually passed to the switch to allow the redirection to take place.

The technique uses the VPI and VCI fields within the cell header. When a cell goes into a switch, the contents of these fields is used to configure the switch. The VPI is used for a VP switch and the VCI for a VC switch. When the cell passes through the switch, the appropriate VPI or VCI is updated with the routing information for the next switch. When the cell arrives at the next switch, it has the control information which is extracted and this switch updates the VPI/VCI field for the next switch from its routing information. This process repeats as the cell goes through the network until it reaches its destination.

When a user makes a connection, each switch in the route is given the routing information that creates the connection route through the network using virtual paths and virtual connections. The paths and connections are described as virtual because they may not take the same physical route or configuration when they are used.

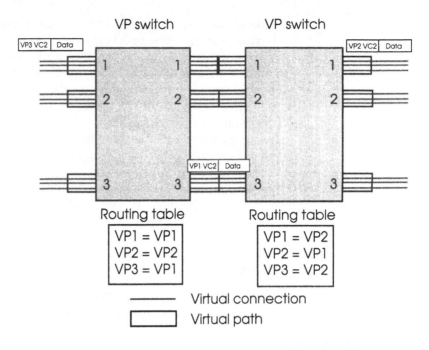

Routing VPs

The routing information is set up by sending cells that contain the routing information through the network. The payload is designated as connection associated layer management information.

The VPI field is eight bits and supports 256 different path connections. The VCI field is sixteen bits and supports over 16,000 routes. In both fields, unused bits are set to zero. If both fields are set to zero, this indicates that the cell is not used and does not carry user data. A field size only defines the number of switching options for each hop between switches. As a result, it can support almost any numbering scheme, such as subscriber telephone numbering or Internet addresses. The address is broken down into as many VPI/VCI fields as needed.

The examples show a point to point connection but multipoint connections are also supported, allowing conferences to be built up and supported.

Making a connection

This concept of supplying the switching information with the data payload to be updated every time the cell goes through the switch has other benefits: because the switching information is established before data is sent, bandwidth through the switch and that part of the network can be allocated to that virtual path and channel.

During the call initiation process, the connection type is requested which defines the connection type and the type of data traffic that is going to be sent. This allows both continuous data, such as video, and burst data, such as data packets, to co-exist.

During the connection process, each switch decodes the destination address for the connection and works out how to route it through its matrix. It will confirm that it has enough bandwidth (virtual channels) to support the level of service that it needs and that there is a virtual path available. In adding the associated routing information to its routing table, it is allocating that bandwidth and effectively committing to provide the quality of service that was requested. If it cannot meet these requirements, the connection cannot be made or must be re-routed. With very intelligent networks and switches, a third option exists that allows bandwidth to be re-allocated from one physical port to another. By doing this, the number of virtual paths and / or channels can be increased and thus remove the bottleneck preventing the connection from being made.

When there is sufficient bandwidth within the network, extra switches can be added to re-route the channels away from the bottlenecks.

AAL service levels

The AAL layer uses cells to process data into a cell based format and to provide information to configure the level of service required.

Processing data

The AAL performs two essential functions for processing data as shown: the higher level protocols present a data unit with a specific format. This data frame is then converted using a convergence sublayer with the addition of a header and trailer that give information on how the data unit should be segmented into cells and reassembled at the destination. The data is then segmented into cells, together with the convergence subsystem information and other management data, and sent through the network.

AAL functionality

The AAL layer as part of the process also defines the level of service that the user wants from the connection. The table shows the four classes supported. For each class, there is an associated AAL.

	Class A	Class B	Class C	Class D
Timing information between source and destination	Required	Required	Not required	Not required
Bit rate characteristics	Constant	Variable	Variable	Variable
Connection mode	Connection orientated	Connection orientated	Connection orientated	Connection less

AAL services

Class A is designed to support constant rate data transmission through a connection, whilst preserving the timing information. This is used for voice transmission. Class B is similar to Class A except that the bit rate is variable and is intended for video / audio data transmission. This is currently being studied. Class C supports a variable bit rate connection with no timing information. The final class is for variable rate, connectionless transmission with no timing information.

There are currently four AALs defined. Originally, there were five with AAL3 existing as a separate entity. It has since been combined with AAL4 to create the AAL3/4 structure. The mapping between classes and AALs is currently defined as:

Class A   AAL 1
Class B   AAL 2
Class C   AAL 3/4 or AAL 5
Class D   AAL 3/4

AAL1

This layer is designed to support voice and requires a constant bit rate, preserved timing information and a connection based mode. It consists of a single layer with a message containing 47 bytes of data from the protocol data unit (PDU). This could be ADPCM or PCM formatted data, depending on the voice encoding actually being used. A header is added with three components.

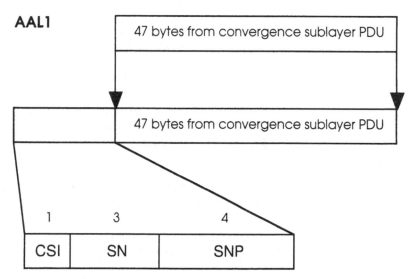

AAL1 services

AAL3/4

This class uses two formats: CPCS and SAR formats. The CPCS or convergence sublayer format takes the data from the protocol data unit or PDU. This can be of variable length and is padded to ensure that it can be divided by four. The header and trailer are then added and the complete data stream is then converted to a set of SAR PDU format cells.

AAL 3/4 —CPCS PDU format

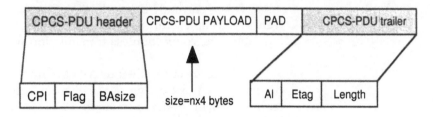

CPI		Connection path identifier : 1 byte
Btag		Beginning tag : 1
BAsize		Buffer allocation size : 2 bytes
PAD		(0...3 bytes)
AL Alignment		1 byte
Etag		End tag : 1
Length		Length of CPCS-PDU payload : 2 octets

AAL 3/4 : SAR-PDU Format

ST	SN	MID	SAR- PDU PAYLOAD	LI	CRC

ST	Segment type : 2 bits
SN	Sequence number : 4 bits
MID	Multiplexing indication : 10 bits
LI	Length Indication : 8 bits
CRC	10 bits

AAL3/4 formats

The SAR (segment and reassemble) PDU consists of chunks of the CPCS PDU with a SAR header and trailer. This information allows the chunks to be collated at the other end and reassembled. The reassembled CPCS PDU then has its header and trailer removed to present the original protocol data to the destination.

AAL 5

This class is for a variable bit rate, connectionless link with no timing information. The process of creating a CPCS PDU from the original PDU, splitting it and appending additional information to create a cell sized SAR PDU is used but it has different formats, compared to AAL3/4.

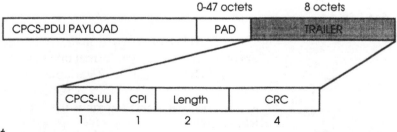

AAL5 CPCS-PDU format

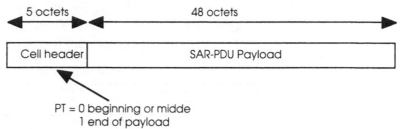

5 octets | 48 octets

Cell header | SAR-PDU Payload

PT = 0 beginning or midde
1 end of payload

AAL 5 SAP-PDU format

Requesting an ATM service

The previous frames provide the mechanism for creating and recreating protocol data units and thus provide the method to transfer data. During the call initialisation, other AAL messages are used to request and set up the service required, including the AAL class that will be used to transfer the data. The two diagrams show the generic format for these messages and the configuration information for an AAL1 service as an example.

			Bits					Octets
8	7	6	5	4	3	2	1	
Subtype identifier								6
1	0	0	0	0	1	0	1	
Subtype								6.1
CBR rate identifier								7
1	0	0	0	0	1	1	0	
CBR rate								7.1
Multiplier identifier								8 (note 1)
1	0	0	0	0	1	1	1	
Multiplier								8.1 (note 1)
Multiplier (continued)								8.2 (note 1)
Source clock frequency recovery method identifier								9
1	0	0	0	1	0	0	0	
Source clock frequency recovery method								9.1
Error correction method identifier								10
1	0	0	0	1	0	0	1	
Error correction method								10.1
Structured data transfer blocksize identifier								11
1	0	0	0	1	0	1	0	
Structured data transfer blocksize								11.1
Structured data transfer blocksize (continued)								11.2
Partially filled cells identifier								12
1	0	0	0	1	0	1	1	
Partially filled cells								12.1

Note 1 -   These octets are only present if octet 7.1 indicates "n x 64 kbit/s" or "n x 8 kbps".

ATM adaptation layer parameters information element for AAL1

Bits								Octets
8	7	6	5	4	3	2	1	
ATM adaptation layer parameters								s
0	1	0	1	1	0	0	0	1
Information element identifier								
1 ext	Coding Standard		IE Instruction Field					2
Length of AAL parameters contents								3
Length of AAL parameters contents (continued)								4
AAL Type								5
Further content depending upon AAL type								6 etc.

ATM adaptation layer parameters
information element

ATM over ADSL

At the beginning of the chapter, it was stated that ATM does not define a physical medium for transferring data but uses existing delivery methods. As such, it does not solve the enigma of how to deliver high bandwidth to subscribers. The advent of ADSL does provide such a solution, and is therefore a natural choice for delivering high bandwidth to the consumer.

The diagram shows the overall structure needed to deliver MPEG2 movies to a consumer using ATM and ADSL. This is a typical multimedia application where the video comes from a video server or a cable television source for a video on demand system.

The audio, video and data streams are multiplexed into a MPEG2 compatible stream. This is then encoded using a H.22x framing to create a PDU. The PDUs use an AAL1 or 5 class and format and these, together with the AAL signalling information, are fed into the ATM layer which uses the ADSL modem to deliver the ATM cells.

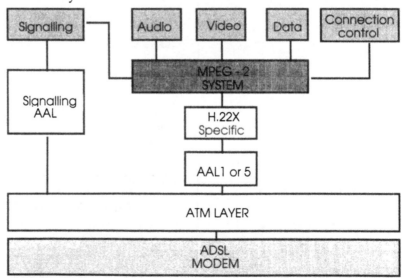

Sending MPEG video over ATM/ADSL

The hardware for such a system is shown in the next diagram. It reflects the same layering as shown in the block diagram. The main difference is that the network end is connected to the video source that provides the MPEG2 encoded data, whilst the other end is connected to a set top box decoder that decodes the MPEG2 streams to provide the subscriber with the movie that has been requested.

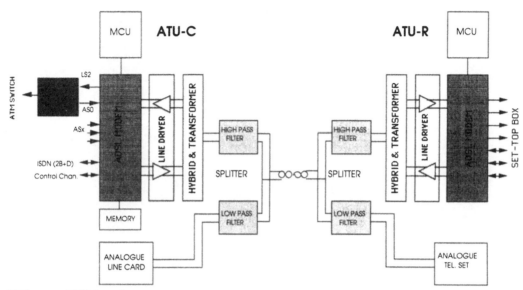

ATM over ADSL hardware

ATM networks

ATM is almost certainly the shape of LANs and WANs to come. It provides the ability to work with a variety of data rates and, by having the same protocol running on a LAN and a WAN, allows the merging of the two to create a seamless integrated network.

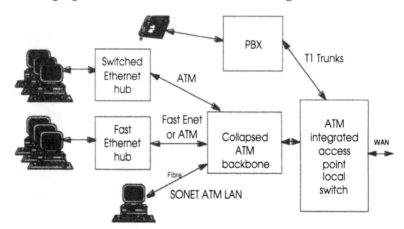

Hybrid ATM network

It is likely that the initial deployments will be in hybrid networks, as shown in the diagram, where parts of the network use ATM and interface using the protocol emulation mode that ATM provides.

While a lot of work still needs to be completed, ATM products have become available. IBM have released a LAN using ATM offering 25 Mbits per second and, with the advent of support chips, the development of other ATM systems will undoubtedly follow.

12 Multimedia conferencing

What is multimedia conferencing?

Multimedia conferencing brings telecommunications and computer technology together into the same product arena. This combination has also brought levels of expectation from the market which differ — and in some cases conflict. Unless the technology can meet theses expectations — and they are not simply concerned with the delivery of audio and video data down a network link to a PC or set top decoder — the uptake and acceptance of multimedia conferencing will be significantly delayed.

Users want interoperability, although the definitions of interoperability vary according to the users' background and experience. Unfortunately, when two separate technologies come together, the expectation is for the best of both worlds and not the worst.

The end-user experience of the telephone is that of a standard service, where the user does not worry about the equipment that the other party has or how the telephone call is actually routed. The connection procedure — dialling the number — is remarkably similar throughout the world and rarely requires resorting to a user manual.

Contrast this to the world of computer technology, where connecting machines together and communication can be a fraught experience. Simply transferring a file can take a major effort, let alone interactive communications. It must be acknowledged that immense improvements have been made to simplify the installation, interoperability and communication of systems — but this is far removed from the standards set by the telecommunications world.

If the market expects interoperability and consistent use, how is the industry going to deliver it?

Achieving interoperability

The key to interoperability within multimedia communications (and particularly conferencing), is the adoption and use of international standards. Whilst this has been standard procedure for the telecommunications industry for almost 100 years, this approach is relatively new to the computer industry, especially the PC market, where standards are based on the dominant products of the time and proprietary definitions. Typically, what happens is that a product dominates (e.g. Windows) and this draws a range of interest in developing further products for this environment. Access is obtained through application programming interfaces, which define how hardware and software can interact with the Windows environment. As new technology becomes available on a computer platform, new APIs (application programming interfaces) are defined to exploit them and to allow application developers to exploit the new facilities.

This development of APIs appears to be similar to the development of the standards used in the telecommunications world. However, these APIs do have restrictions: they tend to be PC specific and assume that any communication is between similar systems. For example, an application that uses the Windows APIs for handling audio and video would need to be rewritten to work on a UNIX workstation or Apple Macintosh. In addition, it cannot be assumed that any data these applications use is also interchangeable.

The problem for application writers is that there are no clear APIs within the computer environment that would allow applications to be able to communicate with any other PC or multimedia device. This is a major barrier which must be overcome as the successful combination of telephony and PC technology requires some method of providing this inter-operability.

Delivering conferencing data

The data delivery system is important for any multimedia system. Its characteristics largely determine the algorithms that can be used and the use that can be made of it.

For most delivery systems, the most important aspect is the bandwidth that can be made available and its real-time characteristics. For local multimedia systems, such as interactive books on CD-ROM, the bandwidth from a CD-ROM player is sufficient to provide the data as fast as the CPU can process it. Even so, there are often blips and jitters when the drive has to change track, creating a momentary delay in the data transfer. To prevent this, faster drives and / or caches are used to buffer the decompression software from these minor discrepancies. The fact that the video playback is delayed while the buffer is filled is of no real consequence within this environment.

If the decompression system cannot cope with the data flow, data is discarded in favour of a slower rate and less processing or, alternatively, the picture size can be reduced. Particularly with software decompression solutions, the more processing power available, the higher the frame rate that can be decompressed and the better the quality seen by the user.

However, if the bandwidth cannot be guaranteed and the data not delivered fast enough, the decompression system may have to drop the frame rate or even skip frames, which degrades the quality of the audio and video.

This problem appears when the compressed data streams are passed over local area networks, such as Ethernet. Ethernet is a non-deterministic network and its performance is greatly affected by loading and traffic density. As a result, there is no guarantee when or if a data packet is going to be delivered. Consider the delivery of a compressed multimedia stream across an Ethernet LAN. When the LAN is lightly loaded, packets are delivered on a regular basis and a high data bandwidth can be achieved. As loading increases and data collisions on the network increase, resulting in many re-transmissions, the data bandwidth falls. To compensate for this, local buffering can be used where the data is buffered on the receiving host to compensate in fluctuations in the data bandwidth. In this way it is possible to transmit audio and video over a LAN. However, the buffering causes the insertion of a latency delay which, whilst it may not be a big issue for some one-way multimedia applications such as video on demand, does pose problems for video conferencing. Conducting a conversation with a half second delay in each speech path dramatically reduces its efficiency, as anyone who has experienced long satellite delays on telephone calls will recognise.

User requirements

Users require a combination of the current experience of both telephony and personal computing, including:

- The ability to make a connection as easily as using the telephone.
- The ability to communicate at different levels with a range of devices.
- The ability to share data and interact dynamically with computers, terminals and other multimedia devices.

Within the PC environment, APIs are now available which provide access to telephony services, such as the TAPI specification for Windows and Apple's Telephone Manager for the Macintosh operating system. Documentation and data sharing is available through technologies such as OLE2 and OpenDoc but, as stated previously, these APIs do assume communication is between similar machines across common networks.

To provide complete interoperability across these diverse platforms, there are two possible solutions: the first is to ensure that each terminal or node uses and provides the same environment. The second is to allow diversity by standardising the communication between the nodes. Given the diversity of multimedia conferencing terminals and network connections, it is the second approach that will provide the only viable method of achieving interoperability similar to that experienced with the telephone network.

Multimedia conferencing — essentially the ability to interactively communicate using audio, video and information data — is thus probably the most technically challenging application for multimedia technology. It has to provide good quality video and audio using appreciably less data bandwidth than that available from CD-ROM or LANs, whilst still synchronising the audio and video to give lip sync without inserting a delay that would make communication difficult or less efficient. In addition, it has to have all the interoperability offered by today's telephone system.

It is interoperability that is the key to understanding and solving these issues. It was recognised by the telecommunications companies (and in particular BT), that for multimedia conferencing to be successful, it had to be as simple as using a telephone, and work with different systems with many different capabilities. Without this framework to guarantee interoperability, it was unlikely that the investment needed to implement the technology would be made — investment essential to generate the low cost silicon to enable its adoption — and to persuade the user to invest in the systems without the spectre of another Betamax-VHS video standards war.

The result of this international co-operation was the H.320 family of standards to provide video conferencing compatibility between any H.320 compliant terminal, irrespective of whether it is PC based, a standalone video conferencing unit or video telephone.

The H.320 multimedia conferencing standards

H.320 defines narrow band (<2 Mbits per second) visual telephone systems and terminals, covering methods of video conferencing, audio, video, graphics and multipoint conferencing between many users. The standard calls on other standards to provide the overall definition. Its main use is with multimedia conferencing over telephone networks, such as ISDN, with typical bandwidths of 64 to 384 kbits per second.

The H.261 encoding algorithm defines the encoding and decoding algorithms for video at both CIF (352 by 288 pixels) and QCIF (176 by 144 pixels) picture sizes. It has a lot of similarity with MPEG techniques: it uses RLE encoding, DCT and motion estimation to provide good quality video using the relative low bandwidths. Another important aspect of the standard is its interactive characteristics and the ability to allow two way interactive communication as opposed to the one way transfer experienced in many multimedia applications.

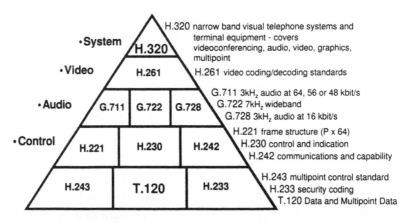

H. series standards

The audio component is encoded using one of three encoding algorithms: G.711, G.722 or G.728. Both G.711 and G.728 provide a 3 kHz bandwidth (the same as provided with an analogue telephone) but use different amounts of data bandwidth. G.722 provides an audio bandwidth of 7 kHz.

The next layer provides the framing, control and capability functions. H.221 multiplexes the video and audio data streams and provides the data channels. H.230 and H.242 provide the control mechanisms and capability exchange protocol. When two H.320 systems start to communicate, they exchange capabilities which define what each party can support, in terms of data bandwidth, picture size (QCIF or CIF), audio encoding, and so on. This capability exchange allows the systems to determine common capabilities which can be used to ensure interoperability. This support also allows very dissimilar systems to communicate. For example, an H.320 system can communicate with an ISDN audio telephone or an H.320 audio-graphics system or even to an analogue telephone. The capability exchange, aided by transparent translation facilities provided by the network, ensures interoperability between H.320 systems and existing telephones and other network devices.

The bottom layer provides additional support for multipoint conferencing, optional security coding and the T.120 series standards.

H.320 overview

H.320 is often used as a blanket reference for video conferencing systems that use all the underlying standards for data and video compression, control, and so on. The H.320 document itself is an umbrella document which describes where and how the other standards fit in and are used. It specifies how continuous and real-time audio and video data can be transmitted and received over a number of 64 kbits per second channels, such as ISDN B channels. It also includes support for the transmission of data, such as still pictures, documents, shared facilities and so on. Also included in the document are the control mechanisms for such calls and how a terminal can add or remove channels whilst maintaining a communications channel.

ISDN call set-up

With the H.320 standard using multiple ISDN basic rate B channels to provide the basic data delivery (this is how most implementations communicate today, but the standard supports multiple channels up to 2 Mbits per second) the key operation is the set-up and control of ISDN channels. This has already been defined using the I.400 series protocols and these are simply referenced within the specification.

ISDN call types

Several ISDN call modes can be specified when making an ISDN call and asking for a channel. Of the set described, the first three are the most important, as most H.320 terminals support at least two out of the three to allow the terminal to communicate with a normal telephone system or fax/modem. The modes are as follows:

Voice   Regular 3.1 kHz analogue voice service. Bit integrity is not assured. This is normally used for a telephone call to another audio only telephone. The bit integrity aspect is important as it effectively excludes data or fax transmission or the use of a modem. This requires that the handset meets all the normal requirements (such as delay) that a normal handset must meet.

Speech   G.711 speech transmission on the call. This is the equivalent of a normal analogue telephone call and does allow the use of a modem or fax. This requires that the handset meets all the normal requirements of a normal handset.

Multiuse   As defined by ISDN.

Data   Unrestricted data transfer. The data rate is specified separately. This is the call mode used for an H.320 call. This does not require the handset delay requirements and so on that the previous modes need to be met.

Alternate speech and data
                   The alternate transfer of speech and unrestricted data on a call (ISDN).

Non call-associated signalling
                   This provides a clear signalling path from the application to the service provider.

The handset delay requirements are interesting; to minimise the delay between the speaker saying something and the speech being transmitted down the cable is an essential requirement to keep the conversation intelligible — if you have ever experienced satellite delay on an international call, you will know how difficult it can make a conversation. As a result, to meet telecoms approval to use the normal speech modes, the delay is measured and must be less than 2 ms.

The ISDN call set-up and control have several functions to perform:

• Make initial calls and establish a B channel connection. This is the simple act of dialling a number to make a connection. With multiple channels, multiple separate telephone numbers may be required or multiple calls can be made to a single number as each B channel is required and connected. As a result, the system has to be able to control the multiple calls and detect call connection drop outs and other phenomena associated with connecting a call. However, the individual channels are treated as one data pipe in terms of the H.320 call. This can cause a fairly complex relationship with the controlling software, as will be seen when some of the protocols are described.

- Establish frame synchronisation between audio-visual terminals.

  As previously stated, each B channel is treated as part of an overall data pipe. This requires several levels of synchronisation to ensure that the data from the different channels are correctly received and assembled. It cannot be assumed that the data frames all arrive at the same time — they may have different paths through the network — and thus each terminal must reconstruct the frames from the data received.

  To ensure that this happens, the data channels between the terminals are synchronised so that each terminal knows how to reconstruct the data and therefore its meaning.

- Exchange terminal capabilities.

  There are many different configurations within the H.320 standards concerning the number of data channels that are used, how much bandwidth is allocated to speech, video, data, and so on. To allow a terminal to communicate with any other terminal, a capability exchange is performed where each terminal declares its capabilities and thus allows the calling terminal to define a common configuration that will allow both terminals to communicate.

- Establish mutually compatible communications between terminals (audio, video, data).

  The main negotiation is performed over the number of ISDN B channels that are going to be used. With most implementations, this varies between 1 and 6 B channels. The bandwidth is divided between the speech, video and data by allocating the speech bandwidth first, followed by the data bandwidth and then any bandwidth that is left over is used for video. The allocation can be changed within certain guidelines so that more bandwidth can be given for a data transfer at the expense of video for example.

  Once the negotiation has established the number of B channels that can be supported, the calling terminal will start to make further ISDN calls to connect the channels.

- Add additional channels if required.

  With the addition of each channel, the additional data is added and synchronised with the existing data channel(s) and the bandwidth re-allocated. Whilst it is possible to add B channels, it is also possible to lose B channels and the process of gracefully losing a channel, re-connecting and re-synchronising is therefore covered within the specification.

- Complete call and disconnect additional channels and initial channel.

  Once the call has been finished, each B channel must be individually disconnected and closed down.

H.261 video coding and decoding

The previous section gives an overview of how the H.320 call is set up and controlled. The next sections describe the individual processes in more detail starting with the video coding and decoding processes, which are defined in the H.261 standard. H.320 terminals are often referred as having an H.261 video codec.

Frame organisation

The standard supports two picture sizes with a variable frame rate of up to 30 frames per second. The larger size is called CIF and has pixel size of 352x288. The smaller picture, called QCIF, is a quarter of the CIF size i.e. 176 by 144 pixels.

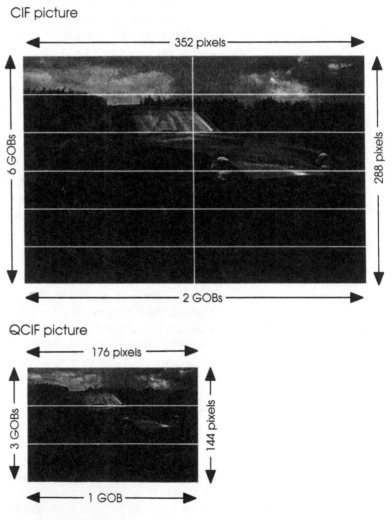

CIF and QCIF pictures

The compression algorithm is like MPEG, in that it uses the DCT transformation coupled with motion compensation and Huffman coding to achieve the compression. Unlike MPEG, it has rate control to allow it to cope with a variable video bandwidth and is designed to cope with bit rates of 40 kbits to 2 Mbits per second. The video format is 4:2:0 luminance and chrominance (YCrCb).

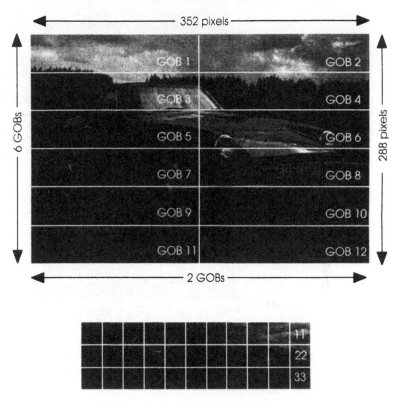

1 GOB = 33 macroblocks

1 macroblock = 16 by 16 pixels (Y) and
                  8 by 8 pixels (Cr and Cb)

H.261 GOB and macroblock definitions

Each video frame is processed by dividing the frame into smaller sections. The first smaller section is called a GOB (group of blocks) and a CIF picture contains 12 GOBs, as shown in the diagram. Each GOB is divided into 33 macroblocks and each macroblock is divided into six blocks. A macroblock consists of four 8 x 8 pixel blocks that form the 16 x 16 pixel matrix, representing the Y luminance information, and two further 8 x 8 pixel blocks for the chrominance information, giving 6 blocks in total and an overall block pixel size of 16 x 16 formed from the Y information with the chrominance information averaged out over the larger block. The numbering sequences for the GOBs, macroblocks and blocks are shown in the diagrams and each block is processed in turn and processed as part of the bitstream. The processing is described in the next section.

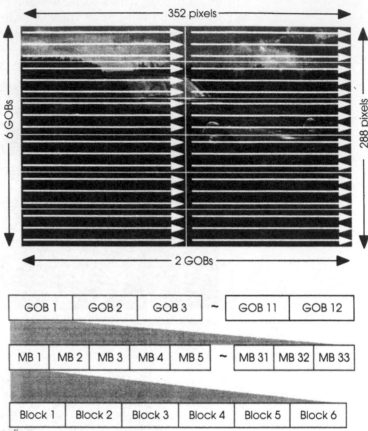

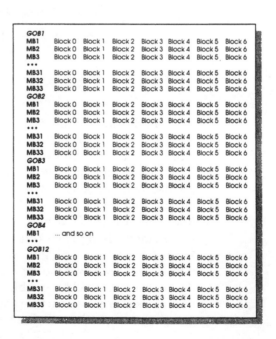

GOB and macroblock coding sequence

H.261 bitstream sequence

H.261 video compression

The basic algorithm is very similar to that used with MPEG. It uses a combination of DCT transformed and compressed frames and vector coding to transmit the data.

The first step is to convert the blocks using the DCT algorithm (the algorithm has been described in the chapters on JPEG and MPEG compression). After conversion, the data is Huffman encoded using a zig-zag path to improve the compression. The DCT coefficients are then sent out to create an intra-coded frame.

There is one important difference with the H.261 method in that the quantisation value used is variable and is determined by the amount of data reduction required to fit the video bandwidth available. With a low bandwidth, the value will be high and the resulting DCT coefficients will be nearly all zero — shown as white in the diagram.

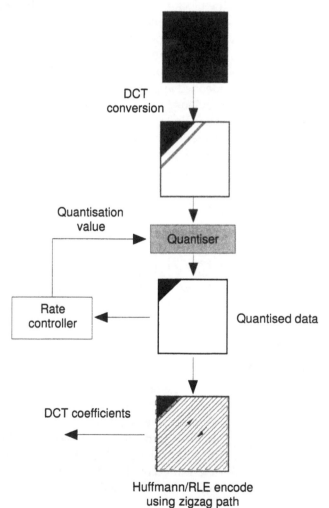

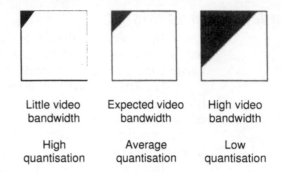

Little video bandwidth	Expected video bandwidth	High video bandwidth
High quantisation	Average quantisation	Low quantisation

Effect on quantisation of video bandwidth

Alternatively, macroblocks can be encoded using motion vectors. These are obtained by searching and comparing the macroblock with other groups within the frame to try and find a good match. If one is found, the vector identifying the pixel offset needed to locate the block is provided instead of the DCT coefficient encoding.

The encoding process uses past frames to encode differences in the same way that MPEG does. However, the H.261 standard restricts the prediction to the previous past frame only. The resulting frame structure consists of intra-coded frames, where the whole frame is encoded and inter-coded frames, where the encoding is based on the previous frame. To prevent errors from accumulating, the standard states that each macroblock must be intra-coded at least every 132 frames. In addition, either terminal can ask for a complete picture update, known as a fast update, which results in an intra-coded frame being sent.

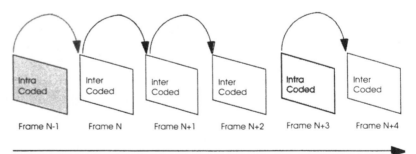

Frame structure

H.261 encoder

The block diagram for an H.261 encoder is shown. It consists of the path already described with additional components needed to use vector compensation and estimation. If the frame is being encoded as an intra-frame, the macroblocks go through the top level of processing and undergo a DCT transformation, quantisation using a value determined by the rate control feedback and finally through the zig-zag Huffman encoding to produce the final bitstream output. The bitstream is also duplicated and sent to a decoder that recreates the frame so that the reference frame is available for inter-encoding.

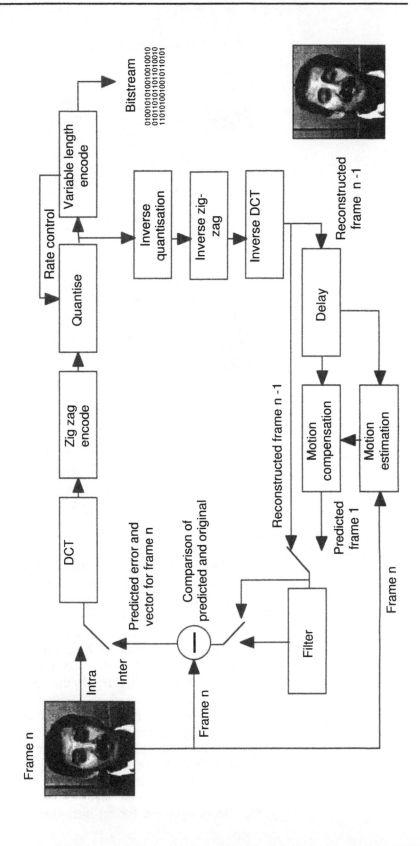

H.261 encoder

Path for intra coded frames

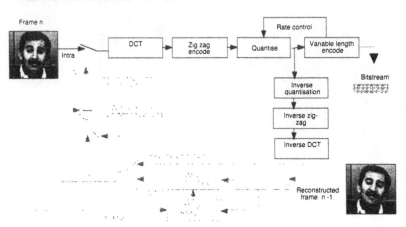

Path for inter coded frames

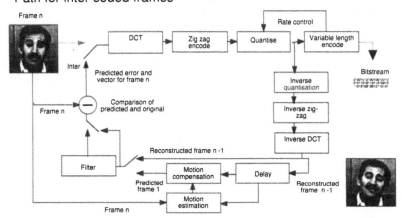

*Encoding paths for inter and
intra coded frames*

The bitstream is decoded so the frame is an exact reproduction of the frame as seen by the receiving terminal at the other end. This is more accurate when comparing macroblocks than using the original source frame. The encoding process is lossy and therefore making the comparison on the original while the receiving terminal uses the decompressed frame as its base can lead to the introduction of artefacts and thus reduce the picture quality.

For an inter-coded frame, the source frame is compared on a macroblock by macroblock basis with the previous frame — hence the need to reconstruct it from the previous bitstreams. The search is performed by shifting the searched macroblock on a pixel by pixel basis to find the best fit. This is similar to the technique used within the MPEG standard described in Chapter 5. If a match can be found that meets the criteria, the vector is sent instead of the complete macroblock. If no match can be found, the macroblock is sent as if it was being encoded for an intra-frame. This information is then processed to produce the bitstream. Headers are inserted as required to allow the decoder to determine the frame type and how the macroblocks were encoded.

H.261 decoder

The decoder path is a lot simpler than the encoder. This has several implications for a terminal which are discussed later.

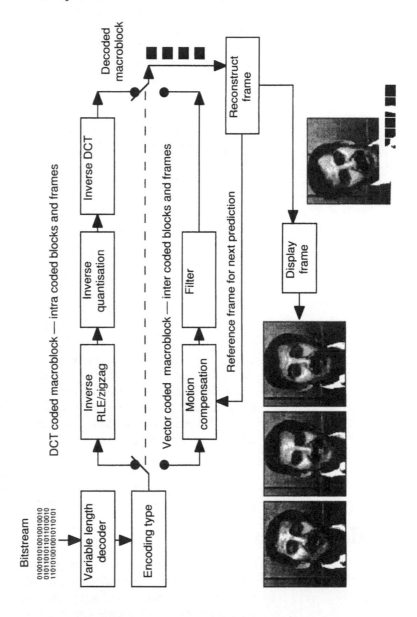

H.261 decoder

The incoming bitstream is received and the encoding type identified. If the macroblock is intra-coded, the DCT-quantisation-encode process is reversed to decode the macroblock. The decoded data is then used to build up the frame. Once the frame has been built, it is available for the display to show. A copy is also stored for use as the reference frame when decoding inter-coded macroblocks and frames.

The inter-coded macroblock is decoded using the vector and reference frame and filtered to improve the appearance before sending the decoded macroblock to be incorporated into the new frame.The complexities of the encoder and decoder are asymmetric: the encoding process is far more complex than the decoding procedure and this leads to an unequal distribution of resources. For a start, the encoder includes a decoder to recreate the reference frames to support inter-coded macroblocks. The macroblock search involves a large number of calculations and adds to the complexity. For video conferencing, both terminals have to be able to encode and decode, and as the user does not see the encoded result sent from his terminal, most compromises are in that area. The strength of a good H.320 terminal is not how good the incoming picture is, but how good the picture is going out.

This is important when it is realised that the standards do not define how to encode the picture but only to decode the picture. As a result, most decoders are reasonable as the standard defines how to exactly process the bitstream. Whilst filtering and other post-decode techniques can improve the appearance of the picture and should not be forgotten, the recovered data is essentially the same and thus its quality is reasonably consistent. Whilst one decoder may support a lower frame rate compared with another, the difference when decoding the same bitstream for individual frames is often very small.

To encode the bitstream is a different matter. Here, quality can be compromised in favour of less processing. For example, the macroblock search area can be reduced so that not all possibilities are checked. The criteria for finding a match can be reduced so that more macroblocks are encoded than inter-coded, thus reducing the bandwidth needed — but reducing picture quality.

This has been one of the problems facing software-only codecs, where software instead of specialised hardware performs the encoding and decoding process as well as the encode. Comparing the incoming and outgoing pictures often reveals where the compromises have been made; they inevitably occur within the decoder. As processing power improves, these differences will disappear.

Video options

Video encoding is primarily concerned with the sending and receiving of interactive video and, in many cases, this is all the video conferencing terminal will support. Larger systems, used in conference suites or as 'rollabouts' can have additional features that can be supported. These options support other cameras and allow other video information to be sent.

Game show format

This is primarily used when large numbers of participants are involved in a conference at one terminal. It allows two CIF pictures to be reduced in height and added together to create two lines of participants, one above the other — hence the game show format term. In the UK, it is often referred to as the 'University Challenge' format after the first popular television game show which pioneered the use of this format.

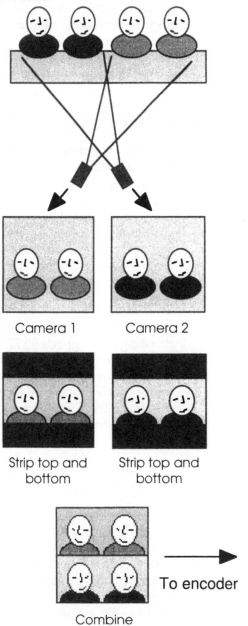

Camera 1   Camera 2

Strip top and   Strip top and
   bottom   bottom

To encoder

Combine
picture

Game show format

Document camera

This is a second camera that can be used to send images of documents or other objects without having to use the conferencing camera. The document image is sent using the same encoding but is identified by a different header and is sent as an intra-coded frame.

Privacy screen

It is mandatory that the video conference starts in audio only mode so that the receiving caller's privacy can be preserved. An electronic privacy screen can be sent instead of the video encoding so that the received video can be seen but the caller will only see the privacy screen. Some countries have taken this to an extreme and require that cameras used with conferencing systems are supplied with a removable lens cover to ensure that privacy is maintained.

G. 7xx audio encoding

So far, we have discussed the video encoding and decoding. Probably the most important medium to be encoded is speech and the H.320 standard defines three speech compression standards to encode and decode audio during an H.320 call.

The standards define the coding to be used by the terminal when it sends audio. It is possible to have asymmetric coding, with one terminal using one standard to send and the other terminal using another — but this is not normally supported or implemented and it is assumed that the same standard is used by both terminals.

G.711

This standard is the equivalent of the a-law/μ-law speech encoding and supports 3.1 kHz audio at 64, 56 or 48 kbits per second. The sampling rate of 8 kHz with an eight bit sample gives the 64 kbits per second bandwidth requirement. The reduced bit rates are achieved by effectively dropping data bits. It is the minimum audio standard supported and is used to establish the H.320 call before moving on to video and more efficient audio algorithms.

G.722

This standard provides higher quality audio with a 7 kHz bandwidth using a 64 kbits per second data bandwidth.

G.728

This standard is similar to the a-law/μ-law speech encoding and supports 3.1 kHz audio but at a greatly reduced bandwidth of 16 kbits per second. The original sampling rate of 8 kHz with an eight bit sample gives a 64 kbits per second bitstream which is compressed using complex digital signal processing techniques to achieve the final 16 kbits per second output. It is probably the preferred audio standard supported because of its low data bandwidth requirement, which means that video can be supported on a single B channel. This is not possible with the G.711 standard, for example. However, the processing is quite intensive and often requires either dedicated hardware or a very fast digital signal processor and software.

H.221 framing structure

Once the audio and video bitstreams have been created, they must be multiplexed together to create a frame that can be sent. The H.221 standard defines the framing structure and how this is achieved.

Octet	Bit 1	Bit 2	Bit 3	Bit 4	Bit 5	Bit 6	Bit 7	Bit 8
Octet 1	Sub #1	Sub #2	Sub #3	Sub #4	Sub #5	Sub #6	Sub #7	FAS code
Octet 2	Sub #1	Sub #2	Sub #3	Sub #4	Sub #5	Sub #6	Sub #7	FAS code
Octet 3	Sub #1	Sub #2	Sub #3	Sub #4	Sub #5	Sub #6	Sub #7	FAS code
Octet 4	Sub #1	Sub #2	Sub #3	Sub #4	Sub #5	Sub #6	Sub #7	FAS code
Octet 5	Sub #1	Sub #2	Sub #3	Sub #4	Sub #5	Sub #6	Sub #7	FAS code
Octet 6	Sub #1	Sub #2	Sub #3	Sub #4	Sub #5	Sub #6	Sub #7	FAS code
Octet 7	Sub #1	Sub #2	Sub #3	Sub #4	Sub #5	Sub #6	Sub #7	FAS code
Octet 8	Sub #1	Sub #2	Sub #3	Sub #4	Sub #5	Sub #6	Sub #7	FAS code
Octet 9	Sub #1	Sub #2	Sub #3	Sub #4	Sub #5	Sub #6	Sub #7	BAS code
Octet 10	Sub #1	Sub #2	Sub #3	Sub #4	Sub #5	Sub #6	Sub #7	BAS code
Octet 11	Sub #1	Sub #2	Sub #3	Sub #4	Sub #5	Sub #6	Sub #7	BAS code
Octet 12	Sub #1	Sub #2	Sub #3	Sub #4	Sub #5	Sub #6	Sub #7	BAS code
Octet 13	Sub #1	Sub #2	Sub #3	Sub #4	Sub #5	Sub #6	Sub #7	BAS code
Octet 14	Sub #1	Sub #2	Sub #3	Sub #4	Sub #5	Sub #6	Sub #7	BAS code
Octet 15	Sub #1	Sub #2	Sub #3	Sub #4	Sub #5	Sub #6	Sub #7	BAS code
Octet 16	Sub #1	Sub #2	Sub #3	Sub #4	Sub #5	Sub #6	Sub #7	BAS code
Octet 17	Sub #1	Sub #2	Sub #3	Sub #4	Sub #5	Sub #6	Sub #7	ECS code
Octet 18	Sub #1	Sub #2	Sub #3	Sub #4	Sub #5	Sub #6	Sub #7	ECS code
Octet 19	Sub #1	Sub #2	Sub #3	Sub #4	Sub #5	Sub #6	Sub #7	ECS code
Octet 20	Sub #1	Sub #2	Sub #3	Sub #4	Sub #5	Sub #6	Sub #7	ECS code
Octet 21	Sub #1	Sub #2	Sub #3	Sub #4	Sub #5	Sub #6	Sub #7	ECS code
Octet 22	Sub #1	Sub #2	Sub #3	Sub #4	Sub #5	Sub #6	Sub #7	ECS code
Octet 23	Sub #1	Sub #2	Sub #3	Sub #4	Sub #5	Sub #6	Sub #7	ECS code
Octet 24	Sub #1	Sub #2	Sub #3	Sub #4	Sub #5	Sub #6	Sub #7	ECS code
Octet 25	Sub #1	Sub #2	Sub #3	Sub #4	Sub #5	Sub #6	Sub #7	Sub #8
Octet 26	Sub #1	Sub #2	Sub #3	Sub #4	Sub #5	Sub #6	Sub #7	Sub #8
Octet 27	Sub #1	Sub #2	Sub #3	Sub #4	Sub #5	Sub #6	Sub #7	Sub #8
•••								
Octet 78	Sub #1	Sub #2	Sub #3	Sub #4	Sub #5	Sub #6	Sub #7	Sub #8
Octet 79	Sub #1	Sub #2	Sub #3	Sub #4	Sub #5	Sub #6	Sub #7	Sub #8
Octet 80	Sub #1	Sub #2	Sub #3	Sub #4	Sub #5	Sub #6	Sub #7	Sub #8

FAS: Frame alignment signal
BAS: Bit-rate allocation signal
ECS: Encryption control signal

H.221 frame

Each frame, as shown in the diagram, consists of 80 octets of information to create eight sub-channels with each bit within each octet allocated to a sub-channel. These are numbered from 1 to 8, with the first seven channels dedicated to carrying audio or video data. The eighth sub-channel is used not only to carry data but also to carry three different codes: FAS, BAS and ECS. This sub-channel is known as the service channel, SC.

The frame alignment signal (FAS) is used to provide a reference for the decoders that receive the sub-channels. The bitrate allocation signal (BAS) is used to control the bit rate and to provide a command path to allow the terminals to exchange instructions, such as forcing a capability exchange, and so on. BAS codes are covered in more detail later. The final signal, the encryption control signal (ECS) is used to control the encryption facilities that may be provided by the system.

In the US, there is an alternative digital data telecommunications standard called switched 56k, which provides a 56k data channel. To enable H.320 compliant terminals to use the network instead of ISDN with its 64 kbps channel, an alternative frame structure is provided where the service channel occupies bit 7 in the octet and bit 8 is set to 1.

Octet	Bit 1	Bit 2	Bit 3	Bit 4	Bit 5	Bit 6	Bit 7	Bit 8
Octet 1	Sub #1	Sub #2	Sub #3	Sub #4	Sub #5	Sub #6	FAS	1
Octet 2	Sub #1	Sub #2	Sub #3	Sub #4	Sub #5	Sub #6	FAS	1
Octet 3	Sub #1	Sub #2	Sub #3	Sub #4	Sub #5	Sub #6	FAS	1
Octet 4	Sub #1	Sub #2	Sub #3	Sub #4	Sub #5	Sub #6	FAS	1
Octet 5	Sub #1	Sub #2	Sub #3	Sub #4	Sub #5	Sub #6	FAS	1
Octet 6	Sub #1	Sub #2	Sub #3	Sub #4	Sub #5	Sub #6	FAS	1
Octet 7	Sub #1	Sub #2	Sub #3	Sub #4	Sub #5	Sub #6	FAS	1
Octet 8	Sub #1	Sub #2	Sub #3	Sub #4	Sub #5	Sub #6	FAS	1
Octet 9	Sub #1	Sub #2	Sub #3	Sub #4	Sub #5	Sub #6	BAS	1
Octet 10	Sub #1	Sub #2	Sub #3	Sub #4	Sub #5	Sub #6	BAS	1
Octet 11	Sub #1	Sub #2	Sub #3	Sub #4	Sub #5	Sub #6	BAS	1
Octet 12	Sub #1	Sub #2	Sub #3	Sub #4	Sub #5	Sub #6	BAS	1
Octet 13	Sub #1	Sub #2	Sub #3	Sub #4	Sub #5	Sub #6	BAS	1
Octet 14	Sub #1	Sub #2	Sub #3	Sub #4	Sub #5	Sub #6	BAS	1
Octet 15	Sub #1	Sub #2	Sub #3	Sub #4	Sub #5	Sub #6	BAS	1
Octet 16	Sub #1	Sub #2	Sub #3	Sub #4	Sub #5	Sub #6	BAS	1
Octet 17	Sub #1	Sub #2	Sub #3	Sub #4	Sub #5	Sub #6	ECS	1
Octet 18	Sub #1	Sub #2	Sub #3	Sub #4	Sub #5	Sub #6	ECS	1
Octet 19	Sub #1	Sub #2	Sub #3	Sub #4	Sub #5	Sub #6	ECS	1
Octet 20	Sub #1	Sub #2	Sub #3	Sub #4	Sub #5	Sub #6	ECS	1
Octet 21	Sub #1	Sub #2	Sub #3	Sub #4	Sub #5	Sub #6	ECS	1
Octet 22	Sub #1	Sub #2	Sub #3	Sub #4	Sub #5	Sub #6	ECS	1
Octet 23	Sub #1	Sub #2	Sub #3	Sub #4	Sub #5	Sub #6	ECS	1
Octet 24	Sub #1	Sub #2	Sub #3	Sub #4	Sub #5	Sub #6	ECS	1
Octet 25	Sub #1	Sub #2	Sub #3	Sub #4	Sub #5	Sub #6	Sub #7	1
Octet 26	Sub #1	Sub #2	Sub #3	Sub #4	Sub #5	Sub #6	Sub #7	1
Octet 27	Sub #1	Sub #2	Sub #3	Sub #4	Sub #5	Sub #6	Sub #7	1
...								
Octet 78	Sub #1	Sub #2	Sub #3	Sub #4	Sub #5	Sub #6	Sub #7	1
Octet 79	Sub #1	Sub #2	Sub #3	Sub #4	Sub #5	Sub #6	Sub #7	1
Octet 80	Sub #1	Sub #2	Sub #3	Sub #4	Sub #5	Sub #6	Sub #7	1

FAS: Frame alignment signal
BAS: Bit-rate allocation signal
ECS: Encryption control signal

H.221 frame for switched 56k networks

1B audio visual call

Given the basic frame structure shown above, the data is sent as frames over a single B channel using the following timings:

- 1 octet is sent every 125 μseconds.
- 1 frame comprising 80 octets is sent every 10 ms.

When the call is initially connected, the audio mode used is G.711, so all the available data bits in the frame are allocated to carry the G.711 encoded audio data. By switching the audio encoding to G.728, the audio data only takes 16 kbits and therefore the remaining data can be allocated to the video data.

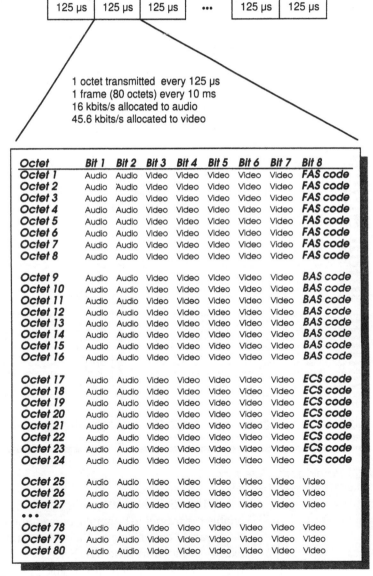

Octet	Bit 1	Bit 2	Bit 3	Bit 4	Bit 5	Bit 6	Bit 7	Bit 8
Octet 1	Audio	Audio	Video	Video	Video	Video	Video	**FAS code**
Octet 2	Audio	Audio	Video	Video	Video	Video	Video	**FAS code**
Octet 3	Audio	Audio	Video	Video	Video	Video	Video	**FAS code**
Octet 4	Audio	Audio	Video	Video	Video	Video	Video	**FAS code**
Octet 5	Audio	Audio	Video	Video	Video	Video	Video	**FAS code**
Octet 6	Audio	Audio	Video	Video	Video	Video	Video	**FAS code**
Octet 7	Audio	Audio	Video	Video	Video	Video	Video	**FAS code**
Octet 8	Audio	Audio	Video	Video	Video	Video	Video	**FAS code**
Octet 9	Audio	Audio	Video	Video	Video	Video	Video	**BAS code**
Octet 10	Audio	Audio	Video	Video	Video	Video	Video	**BAS code**
Octet 11	Audio	Audio	Video	Video	Video	Video	Video	**BAS code**
Octet 12	Audio	Audio	Video	Video	Video	Video	Video	**BAS code**
Octet 13	Audio	Audio	Video	Video	Video	Video	Video	**BAS code**
Octet 14	Audio	Audio	Video	Video	Video	Video	Video	**BAS code**
Octet 15	Audio	Audio	Video	Video	Video	Video	Video	**BAS code**
Octet 16	Audio	Audio	Video	Video	Video	Video	Video	**BAS code**
Octet 17	Audio	Audio	Video	Video	Video	Video	Video	**ECS code**
Octet 18	Audio	Audio	Video	Video	Video	Video	Video	**ECS code**
Octet 19	Audio	Audio	Video	Video	Video	Video	Video	**ECS code**
Octet 20	Audio	Audio	Video	Video	Video	Video	Video	**ECS code**
Octet 21	Audio	Audio	Video	Video	Video	Video	Video	**ECS code**
Octet 22	Audio	Audio	Video	Video	Video	Video	Video	**ECS code**
Octet 23	Audio	Audio	Video	Video	Video	Video	Video	**ECS code**
Octet 24	Audio	Audio	Video	Video	Video	Video	Video	**ECS code**
Octet 25	Audio	Audio	Video	Video	Video	Video	Video	Video
Octet 26	Audio	Audio	Video	Video	Video	Video	Video	Video
Octet 27	Audio	Audio	Video	Video	Video	Video	Video	Video
•••								
Octet 78	Audio	Audio	Video	Video	Video	Video	Video	Video
Octet 79	Audio	Audio	Video	Video	Video	Video	Video	Video
Octet 80	Audio	Audio	Video	Video	Video	Video	Video	Video

H.221 framing structure for a 1B call

The physical allocation scheme allocates data on a sub-channel basis, as shown in the diagram. The first two sub-channels are designated audio and the remaining sub-channels are given to the video bitstream. Note that as a result, the individual octets that are transmitted are thus allocated on a bit basis.

2B audio visual call

For a 2B call, the basic principles still hold — except that the sub-channel allocation uses the additional channel entirely for video as shown in the diagram. In the example shown, the encryption is not supported — thus the ECS bits within the service channel are allocated to something more useful and carry additional video information.

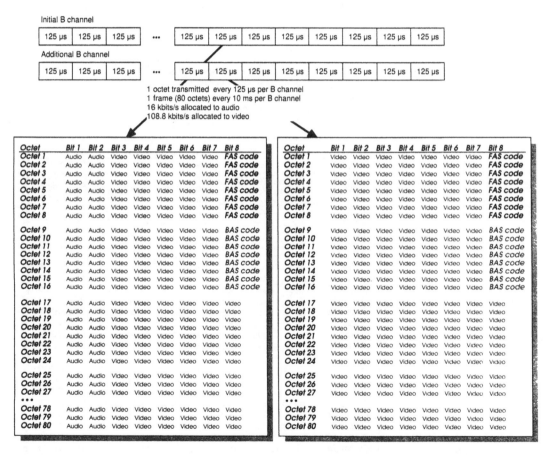

H.221 framing structure for a 2B
call with ECS switched off

In both B channel frames, the FAS and BAS codes are carried to enable the frames to be synchronised and tagged. The frames come in on separate channels and therefore cannot be guaranteed to arrive exactly at the same time. The FAS code allows the components of the frame to be combined correctly and this synchronises the information coming in.

Multiframes and sub-multiframes

The frame structures are combined to create sub-multiframes; these, in turn, are combined to create a multiframe.

A sub-multiframe comprises two frames from each B channel used in the call. It can, therefore, contain a maximum of 12 frames. The two frames are identified as the even and odd frames and carry between them an expanded FAS code that identifies the frames within the multiframe.

Eight sub-multiframes from each channel can be combined to create a multiframe. Each multiframe is numbered and the channel number encoded using bits in the FAS code across all of the frames. This numbering enables the frames to be aligned and checked for correct ordering.

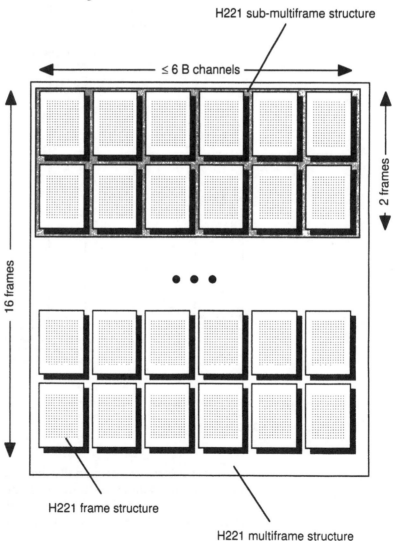

H.221 multiframe and sub-multiframe structures

Controlling alignment

The frames and sub-multiframes that form the multiframe are aligned and synchronised by extracting the FAS codes from each frame. These codes are gathered and placed into a matrix similar to the one shown, where the information that identifies the multiframe and channel numbering can be extracted.

The first check involves the use of the A bit and confirms that a complete sub-multiframe has been received. The N bits provide the multiframe number and are used to combine the sub-multiframes from the various B channels which are feeding the complete multiframe structure. The L bits define which B channel is used to supply the frames and thus determine the source of the frame. The TEA bit is an alarm signal and each frame has the ability to perform CRC checking as indicated by the E bit. The CRC information is stored in the C1:C4 bits within each frame.

Bits 1 to 8 of the service Channel in every frame

Sub-MF	Frame	1	2	3	4	5	6	7	8
SMF1	0 1	N1 0	0 1	0 A	1 E	1 C1	0 C2	1 C3	1 C4
SMF2	2 3	N2 0	0 1	0 A	1 E	1 C1	0 C2	1 C3	1 C4
SMF3	4 5	N3 1	0 1	0 A	1 E	1 C1	0 C2	1 C3	1 C4
SMF4	6 7	N4 0	0 1	0 A	1 E	1 C1	0 C2	1 C3	1 C4
SMF5	8 9	N5 1	0 1	0 A	1 E	1 C1	0 C2	1 C3	1 C4
SMF6	10 11	L1 1	0 1	0 A	1 E	1 C1	0 C2	1 C3	1 C4
SMF7	12 13	L2 L3	0 1	0 A	1 E	1 C1	0 C2	1 C3	1 C4
SMF8	14 15	TEA R	0 1	0 A	1 E	1 C1	0 C2	1 C3	1 C4

Multiframe

N1 to N5 are a five bit multiframe number.
L1 to L3 are a three bit channel number.
TEA is a terminal equipment alarm signal.
R is a reserved bit.
A is the alignment bit(A-bit).
E controls the use of CRC checking for each frame.
C1 to C4 are CRC bits for the frame.

H.221 multiframe FAS code information

This information, especially the N and L bits, is used to determine the source channel and multiframe destinations that the incoming frames are intended for. In this way, the frames can be synchronised and aligned correctly so that the audio and video data can be extracted without errors. Without this information, the wrong data could be placed in a multiframe and corrupt it.

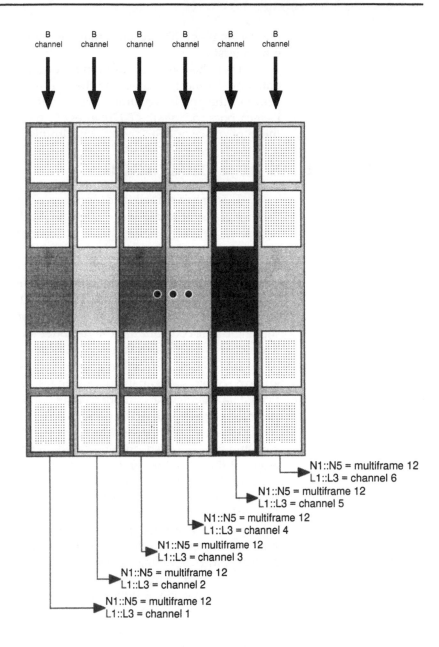

Channel and multiframe information

HSD and LSD data channels

The standard also supports the transmission of data through the high speed data (HSD) and low speed data (LSD) channels. Unfortunately the standard simply defines the channel and not how to use the data. As a result, many H.320 terminals that will happily have audio-visual conversations cannot pass data because of implementation differences. This is analogous to defining an alphabet but not the language. The result is that the words can be transferred but not understood.

There has been an addition to the standard to support a new data transfer mechanism called the multi-layer protocol, MLP. This prevents the problem of unintelligible data communication through its incorporation into the T.120 series of standards. As a consequence, the importance of HSD and LSD is diminishing.

H.230 control and indication

This standard is entitled 'Frame-synchronous control and indication signals for audio-visual systems'. Its function is to define the signalling between the terminals. It incorporates other standards, such as the Q series for establishing ISDN channels, and defines the implementation of the MLP channel for fast data transfer.

It also defines the BAS code sequences used to adjust the characteristics of the call and allow other information to be transferred. Several types of BAS codes exist: the most frequently used is the single byte code but multi-byte codes are also provided to allow a terminal to define its own commands and send other information.

BAS codes are used to change bandwidth allocation, switch audio on or off, start capability exchanges, set data rates for LSD and HSD data channels, add more channels, and so on. They are essentially the control path of the protocols.

When a BAS code is sent to instruct the other terminal to do something, it is based on a declared capability and therefore the assumption is that the command is valid and that the other terminal can execute it. In many cases, the command simply informs the other terminal that the multiframe format is going to change in some way: the audio encoding will change from G.711 to G.722, for example. The other party has a limited time in which to prepare for the change. This real-time requirement can cause problems in non real-time systems, such as IBM PCs. Changing from G.711 to G.728 requires a dramatic increase in processing power — some 40-50 MIPs of fast signal processing power is a typical figure — which is assumed by the capability exchange and declaration to be available. The 7-10 ms available to change the system to respond to the new data structures and decoding is not enough time to download the decoding algorithm from disk, for example. As a result, most implementations are standalone and do not allow the resources to be allocated to other functions. They are essentially reserved for when they are declared and are ready to come into play when the BAS code is received.

The standard states that the interpretation of the BAS codes is left to the terminal implementation. This has led to some obscure incompatibilities between H.320 systems from different suppliers. As a result, an industry group called Versit, whose members include IBM, Apple and Siemens, has released a document that defines how the BAS codes should be interpreted and what the corresponding responses should be. In addition, the IMTC organisation has performed many compatibility trials to identify these oddities and suggest remedies.

H.242 communications and capability

This standard defines the signalling procedures to enable capability exchange, mode switching and frame reinstatement. It uses the commands and signals defined in H.230 and includes them in a protocol to achieve the required functions.

Capability exchange

Capability exchange is an essential part of the H.320 standard. Terminals declare what they can support by exchanging capability tables until agreement on a common set is reached. For example, terminal A may be able to support CIF or QCIF pictures at 30 frames a second using 1 to 6 B channels and all three G.7xx audio coding standards. Terminal B may be able to support QCIF pictures with only 2 B channels and with only the G.711 audio coding. The resultant capabilities that are common are QCIF with 2 B channels and G.711 audio coding. Capability tables are quite large and cover nearly every aspect of the communications, whether it is audio, video or data. The different data rates that can be supported for data channels are included, as well as support for document cameras, game format pictures, and so on.

This technique is also used in many error recovery strategies. If something goes wrong, forcing a capability exchange will bring the system down to a known standard — G.711 audio only — and through the capability exchange allow facilities to be removed.

H.233 security coding (encryption)

This standard defines a method of encrypting the bitstream for enhanced security. The standard is a little light-weight, in that it defines the mechanism and not the implementation. As a result, its adoption has been slow.

H.243 multipoint control standard

Although most video conferences are point to point and only involve two parties, there are times when it is important to bridge several terminals together to create a multipoint conference involving many terminals. To do this, the terminals are connected to a multipoint control unit (MCU) which acts like an audio bridge and distributes the audio, video and data to the participants. This creates several requirements for additional control. For example, which of the participants provides the circulated video? Does the chairman need to see all the participants so that he can select one face for the others to see? These issues are quite immense and are being addressed by the T.120 series standards.

The multipoint control unit provides the basic ability to allow a multipoint control unit to route the selected video and audio to the participants within the conference. The H.243 standard defines the control.

T.120 data and multipoint data

Whilst the H.320 standards provide interoperability for audio video conferencing and provision for data exchange, application sharing, multi-point conferencing and data distribution are defined in another set of standards — the T.120 series.

The T.120 series is a means of communicating all forms of telematics/data between two or more (to very many) multimedia terminals and of managing such communications; it can also manage real-time speech and video where signals are transmitted on channels separate from that carrying the T.120 protocol. It addresses several issues:

- It defines the protocols to be used within the MLP data channel of the H.320 bit stream.
- It provides a multipoint communication service for handling and controlling multi-way conferences.
- It defines a generic conference control layer, which includes the conference set-up and tear down.
- It provides an API for certain application activities such as binary file transfer, chalkboard, and so on.

The motivation behind the T.120 series is to provide interoperability between any terminal capable of supporting audio graphics communication, including an H.320 video telephone, without any participant assuming any prior knowledge about the other system. For example, a PC based video telephone could video and graphically conference with a standalone video terminal. The H.320 protocols would ensure interoperability between them concerning the video and audio whilst the T.120 series specifications would ensure that both could understand how to create and use a common chalkboard for example or exchange data and files. Neither party would necessarily need to have the same application: the T.120 series ensures interoperability by defining the communication and data transfer and extends this up to the application layer by defining how mouse movements are transferred, and so on. The standard application APIs should be seen as interoperability interfaces as opposed to replacement application interfaces.

The standards also cater for network support for fast and efficient distribution of the data stream within a multipoint conference. In this way, an application that uses these standards can work with any other T.120 series compliant terminal, irrespective of its configuration or construction.

It could be argued that these standards are unnecessary because they are already provided within the LAN environment used with PCs today. The problem is that whilst this is true for like systems e.g. Windows PC to Windows PC running the same network operating system software, this level of interoperability does not extend beyond like systems. In addition, although LANs can address many of the data distribution issues, they start to struggle with some of the more advanced multipoint scenarios that are being envisaged. Teacher-pupil networks with shared interactive chalkboards and audio video links are a good example. If several people are using the chalkboard, how does the network ensure that all changes are seen by all participants in the correct order? How does the network cope with distributing data between an ever changing number of pupils without disturbing those already on the network. It is these types of problems that are addressed and solved by the T.120 series standards.

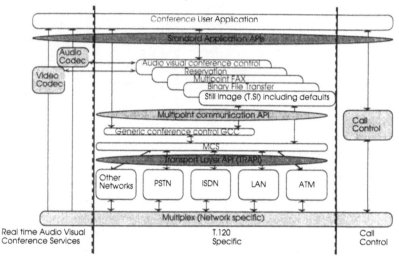

The T.120 standards

H.320 terminals

Throughout this chapter, reference has been made to an H.320 terminal. Several types are available, varying in complexity. The block diagram of an H.320 terminal is shown in the diagram.

Conference suite

The conference suite is a large system that is usually permanently sited in a conference room and consists of multiple cameras and display monitors. It normally supports the game show format, document camera(s) and uses multiple microphones. For data and fax transmission, a PC can normally be connected via a serial port. Such systems normally support up to 6 B channels and even higher transmission rates using primary ISDN links as well as multiple basic rate B channels.

Roll-about

This is similar to a conference suite, except its size is greatly reduced and it normally only supports a single display. It is usually mounted on a trolley that allows the system to be moved from one location to another, hence its name. Again, data and fax transmission is supported using a PC connected via a serial port. The ISDN link is normally based around multiple (1-6) basic rate channels.

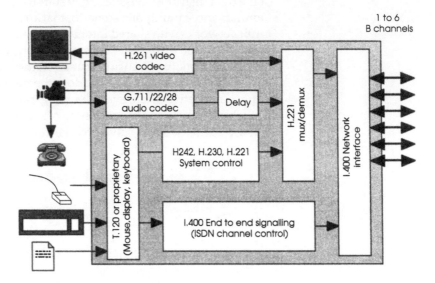

Terminal block diagram

PC based

PC-based systems, such as the VC8000 from BT, consist of a standard PC with a plug-in card that contains the codecs and interfaces to a camera, audio and a handset. The video is displayed within a window on the normal display. Fax and data transmission are usually supported directly within the PC. The communications link is normally based around 2 basic rate B channels. It can support chalkboard and data sharing applications. Some PC implementations can only make H.320 calls and not act as a replacement for an analogue telephone because they do not meet the stringent requirements laid down for a normal analogue handset, in terms of delay and power down mode.

Video telephone

This is a special telephone with a built-in display and camera. It is capable of making H.320 calls as well as calling normal analogue telephones. Such telephones are like the video telephone that has so often appeared in science fiction films. More recent telephones can support data transmission via a serial connection to a PC.

Making an H.320 call

Given that an H.320 terminal has all these components, how is a H.320 video call actually made? The procedures involved are outlined here.

Making the initial call

The first stage is to make the initial call to the other party. Prior to the call, the configuration of the call is set up, usually by taking the default used by the terminal and adding the telephone numbers of the ISDN B channels which correspond to the other party. This makes one large assumption — that the other end is also an H.320 terminal. With the user needing to establish some form of communication, many H.320 terminals will automatically establish if the other party is an audio handset or H.320 terminal and thus achieve some level of communication. The process for doing this is shown in the first flow diagram.

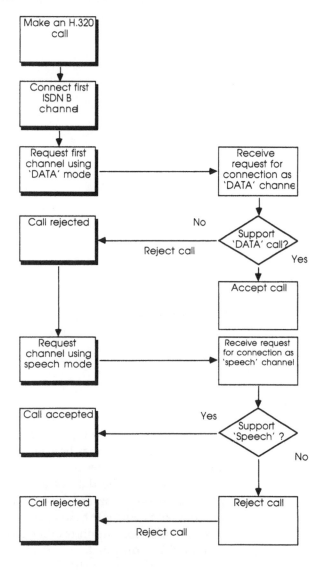

Initial call set-up

The first number is dialled to start the call. (The term 'dialling' is really incorrect as the connection is a digital line — in reality, an

ISDN connection is requested using the destination telephone number.) During this request, the mode must be specified: this is data for an H.320 call. The request is received by the other party, which should check that the requested mode is supported. If it is not, the call should be rejected. The call reject message is received by the originating party — shown by a box with a drop shadow — and could simply say that the call was not accepted. However, in order to check that the other end is some form of interactive communication device, many systems automatically try to establish the call using the speech mode, as if it was an ordinary handset. If the destination can support this mode, at least some form of communication has been created. If not, the originating system should inform the user that the call has been rejected.

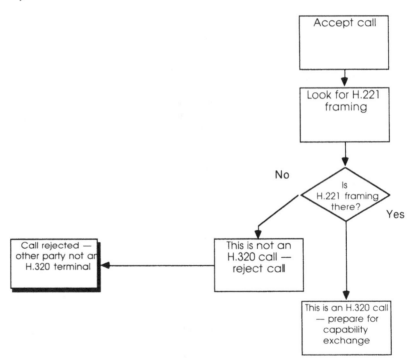

Looking for framing

The second diagram shows the next stage of processing an accepted data call. At this point, the data mode does not confirm that this is an H.320 call. It may be a data call that uses some other protocol and therefore the contents are checked for H.221 framing to confirm and synchronise the data communications. If this framing is not present, the flow diagram shows that the call is rejected on the basis that the call is not H.320. An alternative to this would be to see if the data transmission supports other protocols before deciding that the call should be dropped.

Once the framing has been established (and this must be done by both sides), a capability exchange can be performed to establish the configuration to be used for the communication. As part of this, the number of B channels will be defined. It is up to the originating party to establish the additional channels to reach the required configuration. The call is effectively under control of BAS codes to allocate and configure the H.320 communication and ISDN call control and set up to maintain the data channels.

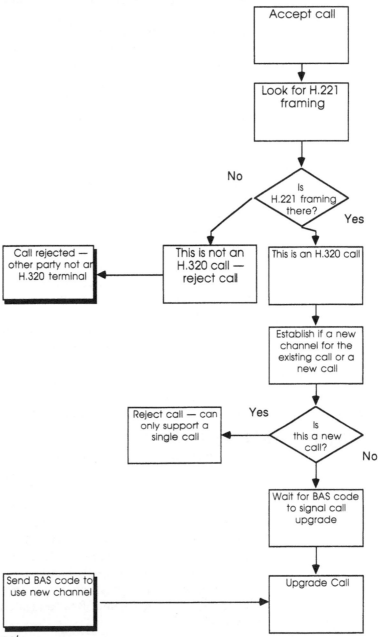

Adding a second channel

At this point, the call is audio only, using G.711 audio. To use a single B channel with video requires the use of G.728 audio to free up bandwidth. This is done by using BAS codes to change the configuration, providing it is within the declared capability. In many cases where video is needed, a minimum of a second B channel is used to increase the available bandwidth. The procedure for doing this is outlined in the next flow diagram.

The second B channel is requested by the originating party as before and is accepted by the receiving terminal, which is expecting a call to establish the second B channel. It is interesting to note that the originating call has to know the numbers it must use to get the channels in advance. These may be different or it may be the same number. With multiple subscriber numbering (MSN), the number used may even vary with the ISDN mode required. The second point is that the calls are all made by the originating terminal and therefore even if a 2B or higher capability is declared, the receiving terminal should wait and not add channels itself. This was theoretically possible but the industry has effectively defined that only the originating party should add calls because of the resultant reduction in complexity in controlling such situations.

The second channel is checked for H.221 framing, as before. If this is present, the call still cannot be assumed to be the second channel: it may be a third H.320 terminal trying to establish a call. The terminal must now check that the second channel is part of the first call by examining the FAS codes and thus confirming that the framing is consistent across the two channels. If not, the terminal must terminate one of the calls (usually the second one), thus preserving the original call and allowing the originating terminal to try again.

At this point, additional bandwidth has been added and the receiving terminal must wait until told by BAS codes from the originating terminal how to allocate the bandwidth, to switch video on, and so on. At this point, the H.320 call is running.

At the end of the call, all the B channels that have been used to create the call must be closed down individually.

Losing and replacing B channels

It is possible to lose individual B channels during a call and this is potentially a major problem. If a channel is lost, the easiest remedy is abort the call, close down all channels and re-establish the call. This is inconvenient, to say the least. If a channel is lost, the H.320 terminals are notified and it is possible to reduce the call to a lower number of channels, close down the lost channels and attempt to re-establish them. Simple implementations reduce a 6B call to 3B if channel 4 was lost and close down channels 5 and 6 before trying to reconnect the channels. A more sophisticated approach takes the five channels, renumbers them to create a 5B call and attempts to reconnect the sixth channel.

Conferencing applications

File transfer

File transfer is an important facility, especially if the H.320 terminal is PC based or connected to a PC. However this is another area where the standards are lagging behind the industry, resulting in the use of many different and incompatible implementations.

One approach has been based on the national file transfer standards which were provided by telecommunications companies to support file transfer by ISDN data channels as a way of getting subscribers to switch to ISDN. However, these standards did not provide the level of file transfer that PC users get with LANs and, due to regional differences, have not been widely adopted. As a result, file transfer has been based on proprietary methods and requires the use of similar applications on both the receiving and transmitting ends.

Some applications have implemented the data channels in such a way that they look like a serial port and thus can be used by any file transfer program that uses a modem and supports the modem based protocols. This approach also has the advantage of providing support for the remote access extensions and support provided by many LANs. This allows remote users to log in and access LANs during a telephone call.

Chalkboard

Chalkboard or whiteboard is a frequently used application which allows participants in an H.320 call to see and change data during an audio-visual call. Predominantly used within a PC, a chalkboard application allows a graphic, text or a spreadsheet to be copied to the board and then allows the different participants to move a cursor and draw on the image. Usually a different colour is used for each participant to allow them to be identified. The changes can then be saved locally.The example shows a drawing which has been annotated: the original could have been copied to the chalkboard by the originating caller and the annotations made by the other party.

This type of application uses the data channels within H.320 to communicate the original data, changes and cursor positions between the parties. It often requires the same system to be used by all participants with the same application, and so on. Whilst it is of great benefit, the data format can also be restricting. On simple implementations, the chalkboard is manipulating a bit map image and not the original data in its original format. For graphics, this is not much of a restriction but with text and spreadsheet data, the ideal would be to allow new text or numbers to be provided by the conferencing which could be used to automatically update the original data at the end of the conference. Whilst this can be done with a bit map image, it is not as efficient and requires the new data to be entered by hand. In this way, it is similar to updating a document from hand-written corrections scribbled on a printout.

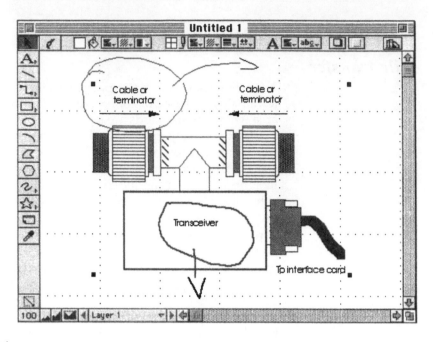

Example chalkboard

Offering this type of facility across different platforms raises several compatibility issues: size of graphics display, colour depth, data formats, and so on, are just part of the problem. As a result, most chalkboards today only work with each other on similar platforms. The T.120 standards address these issues and use capability negotiation to establish common ground. In standardising the information that flows between chalkboard applications and the data format within an H.320 call, many of these restrictions are removed, thus allowing interoperability across different platforms.

Transcoding

Transcoding is the conversion of data from one standard to another. It is often an initial requirement for video conferencing where material stored using MPEG may be converted and transmitted using an H.320 video conference call. This could happen with a video answering machine, where the welcome message is pre-recorded using MPEG1 and the messages are also transcoded using MPEG1. This solves the problem of how to store an H.320 bit stream where the data stream only makes sense if the configuration of the call is known beforehand and thus how the sub-channels have been allocated and to what. Whilst this can be done, it raises problems concerning compatibility with the receiving equipment. With a live call, the capability exchange establishes the communication levels and thus ensures communication between both parties. This is a dynamic resolution of any differences and is made possible because the audio and video are only compressed once the communication has been established.

With a pre-recorded transmission, the encoding has already taken place and thus assumes that the other party can support that configuration. Changing the configuration, renders the pre-recorded data stream obsolete.

To resolve such problems and provide other services, such as video servers, transcoding is used where the H.320 call remains a dynamic communication, except that the audio/video source is obtained from a pre-recorded data stream, such as an MPEG1 stream. The pre-recorded data is decoded to provide video and audio which is fed into the H.320 call as if it was another live source. This process is known as transcoding.

The immediate problem with transcoding is the additional processing that is needed and the time delay involved. With an MPEG 1 bitstream and its future and past motion estimation algorithms that require access to past and future frames, a considerable delay is involved. This is not necessarily a problem with an answering machine: the delay can be tolerated and the transcoding between the H.320 data and the recording MPEG1 format performed after the call has been ended to reduce the processing power needed. If the call involves any interactive activity, e.g. a live voice commentary, with the MPEG1 video describing the scenes, the problems in synchronising the delays become immense: the transcoding takes time and therefore the audio must be delayed to compensate. This incurs another delay in the H.320 process and the result is a call, where the delay and synchronisation reduce the intelligibility of the call.

H.263 video compression standard

As with many standards, no sooner has the first one been developed than the work on the next generation starts. The H series video conferencing standards are no exception to this rule. The H.263 standard is the successor to the H.261 video coding standard for video compression and uses some additional techniques to improve the quality of the video coding and/or the compression ratios. It is designed to work with many different communication channels including local area networks, PSTN modems including cellular connections as well as ISDN.

Its official title is 'Video coding for low bitrate communication' and is aimed at providing video coding for lower bitrates than those provided with ISDN connections and used with the H.261 standard. The techniques that have been used to provide acceptable video quality with a reduced bit rate can also be used to improve the video quality for higher bit rates and thus the standard has rapidly established itself as a replacement for the H.261 specifications. In practice, the H.263 codec is a superset of the H.261 version and thus it is relatively straightforward to design a codec that meets both standards. The H.263 standard forms part of a different video conferencing standard called H.324. Again many H.324 systems are backwardly compatible with H.320 ones.

Supported picture formats

In addition to the CIF and QCIF picture formats that are supported by the H.261 specifications, three other picture formats are added to the standard: sub-QCIF, 4CIF and 16 CIF.

The standard states that not all the formats need to be supported by an encoder or decoder but that there is a minimum set, based on the sub-QCIF and QCIF formats.

Picture format	No of pixels Luminance	No of lines Luminance	No of pixels Chrominance	No of lines Chrominance
sub-QCIF	128	96	64	48
QCIF	176	144	88	72
CIF	352	288	176	144
4CIF	704	576	352	288
16CIF	1408	1152	704	576

H.263 picture formats

Unrestricted Motion Vector mode

This mode is a subtle but powerful change to the motion vector approach. Motion vectors are derived by comparing blocks with others with a search area to find redundancy within the blocks. This means that if a suitable match is found, the vector can be sent to identify where a copy of the matched block should be moved to create the required block. This process is normally limited to the picture boundary and thus prevents looking for matches where the block goes outside the picture edge. If an object is moving out of the picture, then it is logical that a motion vector can be used as the block goes over the edge and leaves the picture as well as when it is completely contained in the picture. If the block overlaps the edge, the outside pixels can be ignored and not displayed.

Search window within the picture

Search window overlaps the picture edge

Urestricted mode search window

This Unrestricted Motion Vector mode allows vectors to point to areas outside of the picture. It is optional and thus provides backward compatibility with the H.261 standard if required. Its use improves the quality when objects move across the picture, especially with the smaller formats.

Syntax-based arithmetic coding mode

This is another optional mode that uses syntax-based arithmetic coding instead of variable length coding. It provides the same quality in terms of the signal to noise ratio and the decoded picture but uses less bits and thus improves the compression ratio.

Advanced prediction mode

This mode uses a technique called overlapped block compensation mode (OBMC) which takes each 16 by 16 pixel luminance block and sub divides it into 4 smaller 8 by 8 pixel blocks that are each used in turn. This method improves the prediction and reduces the number of block type artefacts — areas on the screen which are coarsely defined and where individual blocks stand out — at the expense of higher computation and the use of more bits to encode the multiple vectors.

PB-frames mode

This mode effectively encodes two frames in one and is used to increase the frame rate without dramatically increasing the required bitrate. The name PB-frame is derived from the nomenclature used in the MPEG video coding standard and refers to the ability of using both past and future frames as references for the motion compensation.

A PB-frame contains a P picture which is predicted from the last decoded P picture and a B picture that is predicted from the last decoded P picture and the one currently being decoded. This technique allows additional frames to be encoded without increasing the used bit rate very much. The end result is a higher frame rate which can be beneficial for small format video.

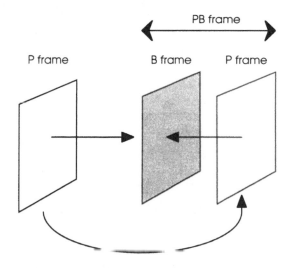

The PB frame

Error handling

Although error handling is specified to be handled by external means — usually an error correction code — optional error coding mechanisms are specified. These are inserted into the framing structures to protect the data contained in the frames.

There are some additional facilities: a decoder can instruct the encoder to help it prevent buffer overflow by intra-coding one or more of its GOBs using certain parameters. This effectively increases the compression, reduces the required bit rate and therefore helps prevent overflow. An alternative strategy for the decoder is to request that only non-empty GOBs are sent. Again this helps reduce the required bit rate. These signals are not sent in-band but by using some other external channel.

13   Digital video broadcasting

Digital video broadcasting (DVB) is the name given to a world-wide initiative to replace the current analogue-based television systems with a digital service. The reasons for the switch over are many and varied:

- Improved quality of service.
- The ability to fit more channels into the existing frequency bandwidth.
- The ability to mix and match services.
- The ability to provide interactive services and national and regional options from a central facility.

The delivery mechanisms

DVB systems will deliver multimedia — audio, video and data — via several systems, depending on the existing networks for analogue systems. Several such systems can be used: satellite, microwave, telephone, cable and terrestrial broadcast.

Satellite

Satellite systems already deliver a large number of channels and are possibly the favoured method of delivering DVB over a wide geographic area.

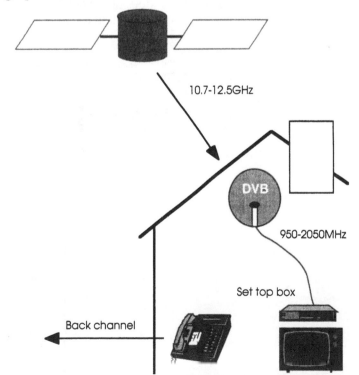

10.7-12.5GHz

DVB

950-2050MHz

Set top box

Back channel

Satellite delivery of DVB

Many existing satellite systems are already capable of digital transmission and all that is needed to start a DVB service is a change to the data uploaded to the satellite and the replacement or adaption of the satellite receiver (the set top box) to enable the decoding of digital transmissions. Using the data compression techniques provided by MPEG2, the adopted standard for DVB compression, a single satellite could provide several hundred television channels.

The problem with a satellite is that it does not provide a back channel within its communications. This means that to support the interactive aspects of the DVB services, or simply to pay for an encrypted service and get the decryption, a second communications link is needed. In most cases this could be a telephone. However, with both the telephone and cable companies working to provide programs in their own right, this is not an ideal situation for a satellite operator.

Microwave

Microwave links are also good providers of high bandwidth communications. They are expensive and rely on a line of sight link. They already form a strategic part of the telephone network and thus have some potential for distributing DVB services, possibly acting as a central reception point for high rise buildings, for example.

Telephone network

The telephone network has the advantage of a built-in back channel but it does not provide the same amount of bandwidth as a satellite system can deliver.

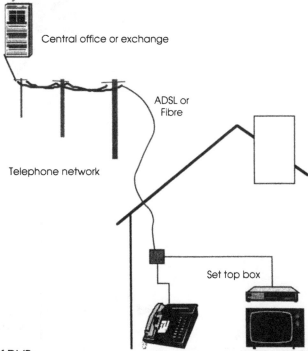

Telephone network delivery of DVB

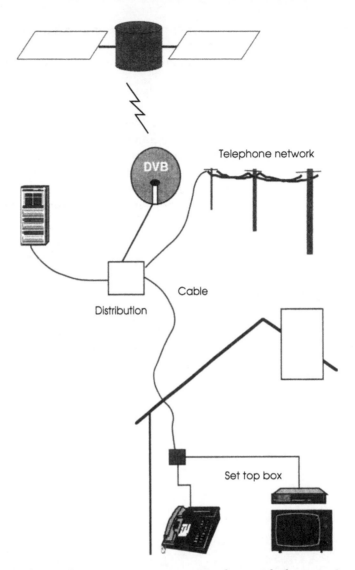

Cable delivery of DVB

This is changing and it is possible that, with the introduction of fibre optic or coax cable instead of twisted pair to the home, that the telephone network will evolve to being the central point for providing multimedia data to the consumer. Again, a set top box would be needed to decode the DVB transmission and would be connected to the network via a splitter which also provides the telephone connection.

With the advent of ADSL, there is enough bandwidth to provide some television programs. It is likely that this service will form the basis of a video server or video-on-demand service, where the consumer uses the back channel (i.e. the telephone) to request a specific video or video information. This can offer the breadth of services that is possible from a satellite — but it relies on a direct connection to each consumer and multiple video sources to support it. Several trials of such services are currently being conducted.

There are problems for telcos in providing this service which go beyond the normal technological issues. For example, BT in the UK is prohibited from providing television programs via its telephone network.

Cable

Cable delivery is possibly the most flexible option as it can combine programs from satellites, video servers and even the telephone network to provide an integrated service. Without the limitations that many telcos have to work with, they can provide a telephone connection service, and thus a back channel, using the telephone or their own cable back to a central point. For new housing developments, this is an ideal method of providing services. For existing services and especially old or remote properties, the cost of installing cable can be expensive and in many cases prohibitive.

Terrestrial broadcast

For countries where the predominant method of delivering television is by radio transmission, moving from an analogue to a digital system provides more bandwidth and thus allows the number of services to increase. In the UK, the expansion of the national television services has been limited by the lack of available bandwidth. With the switch to digital encoding, the number of channels that can be supported is increased and thus further expansion supported.

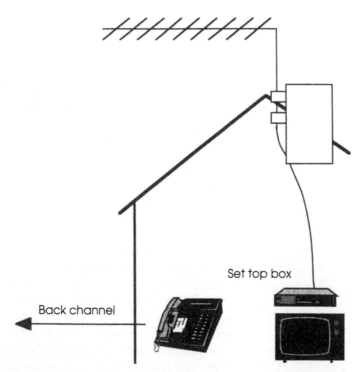

Set top box

Back channel

Terrestrial broadcast

Encoding techniques

The first challenge that has to be overcome before DVB is a reality concerns the modulation/encoding techniques that must be used to get the bandwidth over as small a frequency spread as possible. The normal techniques of amplitude modulation, where a carrier wave's amplitude changes as dictated by the signal wave, and that of frequency modulation, where the carrier wave's frequency is changed as dictated by the signal wave, cannot carry the digital data or data density that is needed. An alternative technique is needed, based on changing the phase relationship of the carrier wave.

Phase shift keying

With phase shift keying, the data is encoded by the phase relationship between the carrier signal and a reference signal of the same frequency. The amount of phase shift determines the data value that has been encoded. To decode the signal, a locally generated reference signal is combined with the incoming carrier. The resulting signal contains peaks and troughs as the carrier moves in and out of phase with the reference signal. The local reference is normally a heavily damped phase lock loop circuit (PLL) that generates a reference signal which matches the frequency and averages out the changing phase relationship to create a good enough copy of the reference to allow decoding.

The term keying is used instead of modulation because the modulation is not analogue where the waveform can vary but is effectively a digital on or off signal. In this way, it is similar to the old telegraph key that was used to send data by Morse code.

Neither the amplitude or the frequency play a part in the decoding and thus provides a good reliable transmission medium that is less susceptible to errors. By adding amplitude modulation to the basic phase encoding, the number of data bits can be increased. This helps achieve the high bit density that is needed to conserve the bandwidth used to transmit the data.

Binary phase shift keying

This encoding method can carry a single binary bit per symbol, where the term 'symbol' is used to define the smallest amount of the carrier wave that can carry data or be modulated. In the diagram, the symbol is two complete cycles of the wave form and each bit is encoded using the two cycles. Depending on the implementation, the symbol's size can be greater or smaller, going down to a single cycle.

The encoding used in the example modulates one data bit by setting the phase shift to zero, and modulates a zero by adding a 180 degree phase shift. When this carrier is combined with a locally generated reference, the phase shifted carrier combines to create an increased peak where there is a 1, and a reduced amplitude where there is a 0. A comparator can be used to extract the digital bit data from these levels.

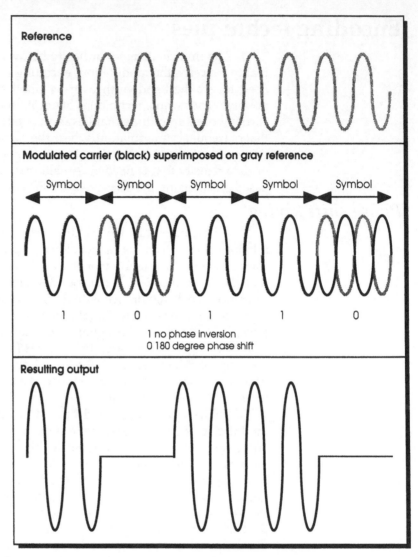

Binary phase shift keying

Quadrature phase shift keying

In the previous example, the phase shift was essentially a single value of 180 degrees. By reducing the phase shift value, more bits of data can be encoded. This is the concept behind quadrature phase shift keying (QPSK). The term quadrature has a dictionary definition of 'measuring the position of something using a reference object which is located at an angle of 90 degrees'. The 90 degrees gives a clue as to how this method works. The phase encoding is performed in steps of 90 degrees. This gives four possible phase shifts. Each symbol can now carry two bits of digital data.

A common way of representing the encoding is by using a diagram that plots amplitude against phase shift. This gives a circular plot, often called a constellation. The encoding defines the possible points which are dependent on the phase relationship and the resultant amplitude when combined with the local reference signal. As the modulated and reference signals are of fixed amplitude, and the modulated carrier can only be out of phase for four defined values, four points are defined on the graph which represent the data. The amplitude axis is often referred to as the Q axis and the phase axis is known as the I axis.

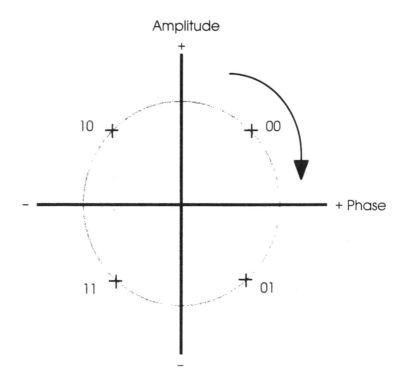

QPSK encoding

The encoding for the bits is known as Gray coding. It has the property of only changing one bit as you move around the values. The advantage of this coding is that each bit of data changed requires either a phase or an amplitude change. There is no case where both bits change as a result of a single change in amplitude or phase. This improves the quality of the encoding and makes it less likely to produce an error.

Quadrature amplitude modulation

Quadrature amplitude modulation (QAM or QUAM) is a derivative of the previous technique. Here, the phase encoding is still used but the amplitude is also modulated to create even more fixed points in the constellation. There are several different implementations which use a different number of amplitude levels.

16 bit QAM   Defines 16 points and effectively encodes 4 bits of data per symbol.

32 bit QAM   Defines 32 points and effectively encodes 5 bits of data per symbol.

64 bit QAM   Defines 64 points and effectively encodes 6 bits of data per symbol.

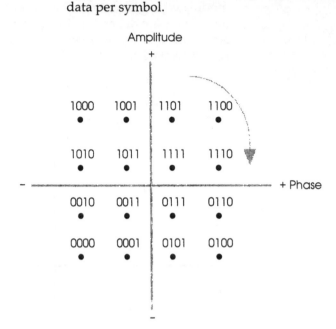

16 bit QAM

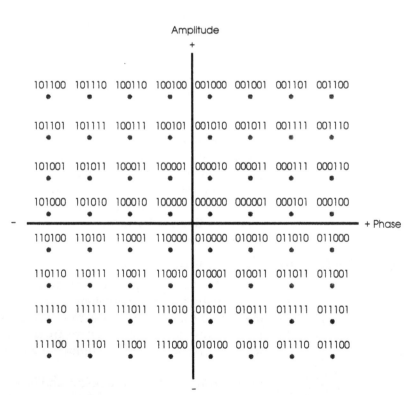

64 bit QAM

The 64 bit QAM signal is created by taking the six bit data and splitting it into two three bit halves. Each half is used to amplitude modulate a carrier into eight discrete levels. The two carriers are 90 degrees out of phase with each other. They are quadrant multiplied together and summed to create the final signal, which translates the two sets of 8 discrete amplitude values into 64 points, each with a unique amplitude and phase relationship.

Typically, a symbol can occupy 1 Hz of bandwidth. This allows a very high bit rate to be transmitted over a small bandwidth, thus achieving high data density.

	Occupied bandwidth (MHz)	Symbol rate (Mbaud)	Bit rate (Mbit/s)
64-QAM	7.92	6.89	41.34
32-QAM	7.96	6.92	34.61
16-QAM	7.86	6.84	27.34

QAM coding and bit rate

For DVB, 64 bit QAM has been chosen as the main encoding technique. Support for both 16 and 32 bit QAM is also included.

In practice, 8 MHz of bandwidth is used to carry a number of 64 bit QAM carriers. Each carrier is allocated 8 kHz, theoretically giving 8,000 carriers. In practice, this is not the case. Only 6,785 carriers are used, due to filtering and channel separation — carriers are spaced out across the bandwidth at set multiples of a frequency. This is usually slightly bigger than the bandwidth required, to compensate for the lack of perfect filters and thus ensure that carriers do not overlap and distort each other.

As the symbols are transmitted using the carriers, there are small time intervals (guard intervals) where the symbol is not transmitted. This sacrifices some bandwidth, as during a guard interval, no symbols or data is sent, but it does prevent interference between the symbols. The combination of these techniques is known as orthogonal frequency division multiplex (OFDM).

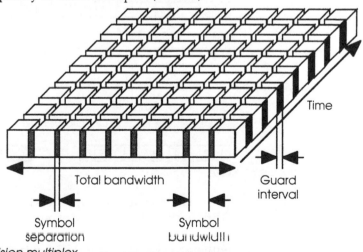

Orthogonal frequency division multiplex

Other coding techniques are used to enable the decoder to recover as much signal as possible. These involve the use of data redundancy and add information to the contents of the packet, as is explained in the next section.

Coping with burst errors

Many of the delivery mechanisms discussed here are not 100% error free. As a result, some of the available bandwidth is used to cope with errors and their correction. This is known as forward error correction (FEC). The basic technique involved uses additional data, often referred to as CRC (cyclical redundancy code), to allow blocks of data to be reconstructed if data is missing. CRC is just one of many techniques available. The actual coding is based on mathematical algorithms and can detect different types of errors. The one used within DVB is Reed Solomon.

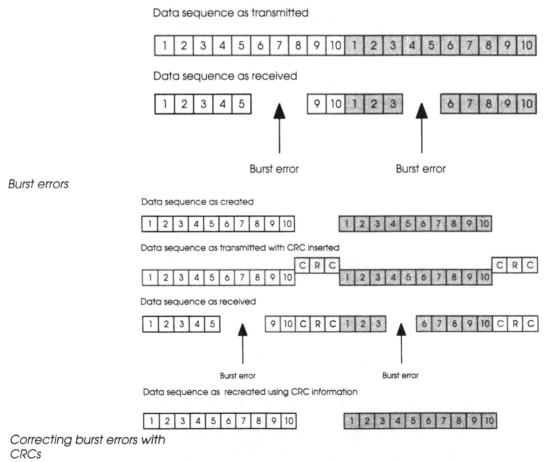

Burst errors

Correcting burst errors with CRCs

The need for this support is to provide the level of service that DVB specifies. Its definition of quasi error free (QEF) means less than one uncorrected error event per transmission hour. The bit error rate within an MPEG2 transport stream is about 1 in 10^{11}. To match the two requirements, error recovery is used at two levels. The first or outer level uses Reed Solomon encoding, as already described. The second uses a technique called convolution.

Convolution

Convolution is defined in the dictionary as 'a twist'. It describes a complex process that takes two sets of data and allows them to be reduced to a single set. The convolution process produces the same end result in spite of errors in the transmitted data. The convolution ratio defines how much redundancy is used. A rate of 1 to 2 has 100% redundancy, whilst smaller ratios have less redundancy and therefore less tolerance to errors. DVB uses a ratio of 7 to 8.

To decode the data and retrieve the original data, the Viterbi algorithm is frequently used. This is very computationally intensive and is either done in hardware or with a fast signal processor chip.

Interleaving

Although the Reed Solomon error correction coding is efficient, it does require a lot of additional correction bits to cope with burst errors within a transmission.

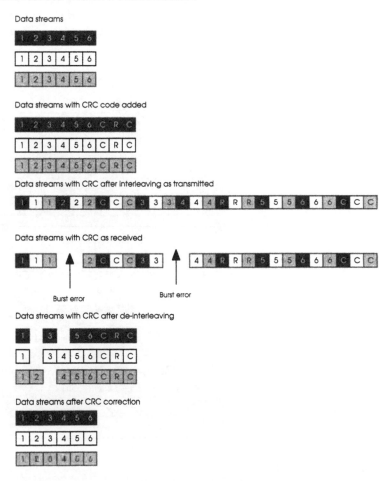

Interleaving

Interleaving is a technique which reduces the size of the required Reed Solomon code by not sending the packets as sequential data. Instead, the data is sent in small blocks, interleaved with other stream information. If a burst error occurs, the lost data is not specific to a particular stream but is divided across several streams. The number of bits per stream lost is therefore less and the size of the CRC code needed reduced, as shown in the picture.

Pseudo random distribution

There is another refinement that is employed by DVB and that is to use a pseudo random number to scatter and distribute data across the spectrum so that a more even distribution is achieved. It is used to address the problem caused by digital data sequences which result in only part of the bandwidth being used with a large amplitude component.

For the best quality of transmission, it is beneficial to spread the data out over the entire available bandwidth so that no particular frequency bears the burden of carrying the data. By using a pseudo random code sequence that both the encoder and decoder understand, the data can be distributed across the frequency spectrum and re-assembled correctly by the decoder. This technique is used within DVB to improve the signal quality. The diagram shows the basic technique.

Before pseudo randomisation

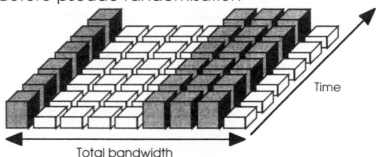

After pseudo randomisation

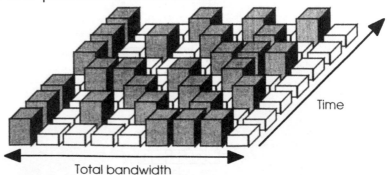

Pseudo randomisation

The three main delivery systems do not use the same path for decoding the data. The processing is dependent on the quality of the transmission path and, as can be seen, varies considerably between cable, satellite and broadcast. In the worst case, 16 bytes are added to a standard MPEG2 transport stream packet to create the error correction codes.

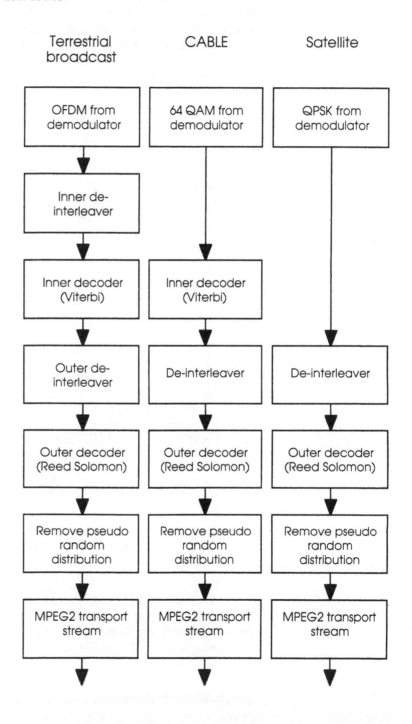

Delivery mechanisms

DVB data streams

The contents of a DVB data stream is based on the MPEG2 standards. It uses the MPEG2 transport mechanism designed for high bandwidth and high error rate transmission, as found with satellite, terrestrial broadcast and cable. This is exactly the requirement for DVB, which multiplexes many MPEG2 compressed programs into a single stream, thus allowing different programs to be chosen from the bitstream (refer to Chapter 5 on how MPEG compression works).

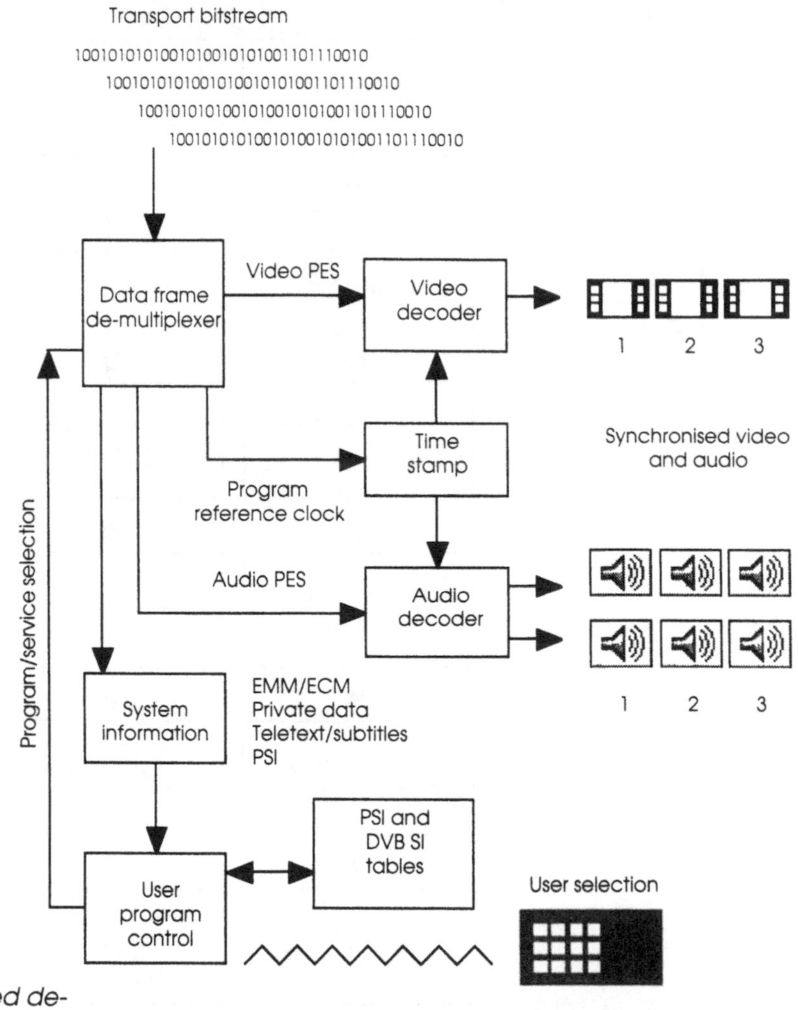

Transport bitstream

100101010100101001010100110111110010
100101010100101001010100110111110010
100101010100101001010100110111110010
100101010100101001010100110111110010

Data frame de-multiplexer

Video PES

Video decoder

1   2   3

Synchronised video and audio

Program reference clock

Time stamp

Audio PES

Audio decoder

1   2   3

Program/service selection

System information

EMM/ECM
Private data
Teletext/subtitles
PSI

PSI and DVB SI tables

User program control

User selection

Transport stream based decoder

DVB is based around the main level and main profile categories. It uses packetised elementary system (PES) elements to construct the payload but, unlike the program stream used with more reliable delivery systems such as CD-ROMs, it uses a fixed size packet of 188 bytes to deliver the data. The packets also contain the program specific information (PSI) used to reconstruct and control the various programs within the stream. This is the basis of the additional services that will be offered through DVB. Each program — video and associated audio and data channels — has its own clock reference.

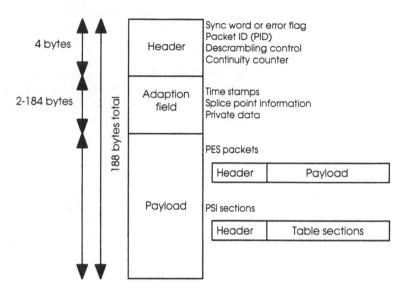

MPEG2 transport stream

Packetised elementary system components

The MPEG2 standards specify the following PES types:

Video   The normal video bitstream. It contains the encoded video information for a specific program.

Audio   The normal audio bitstream. It contains the encoded audio information for a specific program.

Private_1 The first private data channel.

Private_2 The second private data channel.

DSM   These commands provide control of a local digital storage medium, such as a CD-ROM or hard disk.

EMM   Entitlement management message.

ECM   Entitlement control message. This, coupled with the EMM, provides the communication channels for the controlled access to programs either through smart cards or through the use of PIN numbers. For example, a subscriber can ring a supplier, give a credit card number to access a particular program or service and the access code or encryption key can be sent to the decoder via one of these messages.

Teletext   Provides teletext information for a program. This is normally encoded with the analogue signal in analogue systems. With MPEG, this is not possible and the teletext information is coded as an external data channel.

Subtitles	Provides subtitling information for a specific program.
Padding	Needed to ensure that the transport packet is always 188 bytes.
Others	For future expansion.

Program service information

The MPEG2 standard specifies the following PSI tables to store the PES program specific information. DVB has defined a set called service information (SI) which complements the set defined by the MPEG2 standards and provides additional services.

PAT	The program association table maintains the links between the program number and the packet identifiers so that the packetised information can be identified and linked to the other packets to create the programs. Without this information, the decoder cannot cope with multiple programs.
PMT	The program map table is responsible for defining what a program actually is and the services that it includes, e.g. different sub-titling, data and other facilities.
CAT	The conditional access table is used with EMM data to control conditional access to programs. This is used for pay television and similar services. DVB uses two types called simulcrypt (based on smartcard technology) and multicrypt, which currently uses a PCMCIA card.

DVB service information

The SI set of tables extends the MPEG2 PSI to support the needs of the broadcast television industry and complement the PSI. It consists of:

NIT	The network information table contains information about the physical network, such as channel frequencies, satellite transponder number, and so on.
BAT	The bouquet association table contains the information that defines how groups of services are combined into bouquets (see below). The sources of the services need not be delivered by one system. For example, services could be obtained from cable, satellite and terrestrial broadcast.
SDT	The service description table contains information about the services, such as the names of services, providers etc.
RST	The running status table gives the status of an event (running or not running).

EIT	The event information table contains data about current and future events, screen format, type of audio, etc.
TDT	The time and date table gives present time and date.
ST	The stuffing table is used to invalidate existing data and associated entries.

DVB bouquets

The data provided in the DVB service information allows the concept of a service bouquet to be defined. In essence, a bouquet is a collection of services which may originate from more than one network. Each service within the bouquet can potentially supply all the PES components such as video, audio, teletext, sub-titling etc.

The aggregation of the services from different sources into a bouquet is performed by sending information to the bouquet association table, which defines the bouquet and which services are included.

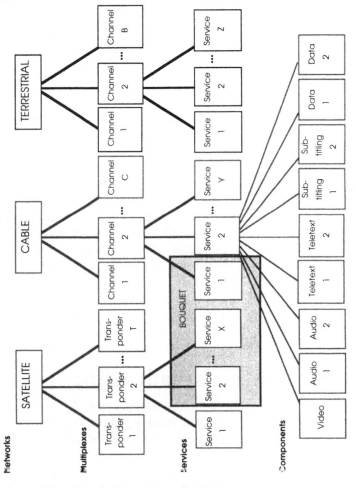

DVB service bouquets

Video servers and video on demand

The terms video on demand (VOD) and video server have become synonymous with multimedia. The basic idea is simple: the user is connected to a video server that lets him choose the video program from a library or database. Once selected and paid for, the audio-video program is transmitted to the user over the connection of choice. For most installations, this would be cable based — either ADSL (telco connection) or coax (cable company).

There is some debate over the level of service. With a true VOD service, the video can be requested at any time. This potentially means that every subscriber would need a video playback unit within the video server to provide this information. Clearly, this is not feasible. By limiting the number of choices, the number of video players is reduced and the programmes simply require distribution to the subscribers who have requested them. This is fine, but for the problem of response time. For true video on demand, the video should start almost immediately — but this again starts to define a system design with a player per subscriber, so that individual start times can be supported.

As a compromise, some video on demand trials have implemented a choice of videos with a set of defined start times, say, every 10 minutes. This reduces the complexity of the system at the expense of the response time. These systems are known as nearly video on demand (NVOD).

Whilst the immediate interest is in an alternative to going to the local video shop and choosing a cassette from their library, this type of technology has other applications. There are huge libraries of video material gathered over the years from newsreels, television programmes and other sources, which form a fascinating library for research and other similar activities. The problem has been in getting access to it. A multimedia communications system, where a user can dial in and access a server and explore the video library over the telephone or from the home has many possibilities. With the ability of PCs to handle audio video data as if it was text, the problem of supplying low cost terminals also disappears.

To confirm that this is a route being actively pursued, many of the database software companies (e.g. Oracle) have developed support for audio and video within their databases so that information can easily be searched for and identified.

Digital audio broadcasting

In September 1995, the world's first digital audio broadcasting (DAB) service was started by the BBC in the London area. It has been made possible by the development of a set of standards for DAB known as Eureka 147. With the standards and technology defined, the system was effectively seeded by the BBC to encourage receivers to be developed. This effort was funded as a public service to break the chicken and egg syndrome — there are no DAB services because there are no DAB radios and there are no DAB radios because there are no services.

The service provides CD quality stereo sound with very rugged and reliable transmission. It does not suffer from the reflections and signal deterioration that FM broadcasts do in cities or in the shadow of large buildings or other objects. Like DVB, it allows more channels to be supported in the same bandwidth — and this again is a major incentive to the adoption of this technology.

The technology has a lot of similarities with DVB, so that there is a common leverage and re-use of investment. It uses MPEG1 transport streams and the audio is encoded using the audio layer II standard. This achieves a compression ratio of 8:1, whilst maintaining the subjective audio quality of CD. This is based around a psycho-acoustic model with 32 sub-band analysis and is explained in more detail in Chapter 5. Typically, 192 kbps of bandwidth is needed to support a stereo channel to this quality, although this can be switched to a lower data rate of 96 kbps, depending on the required audio quality and program content.

Data can also be carried which can be used to provide additional program information and be used to control the services that are provided, including conditional access for pay as you listen services. Like DVB, the transmissions can use many different delivery methods e.g. radio broadcast, satellite and cable.

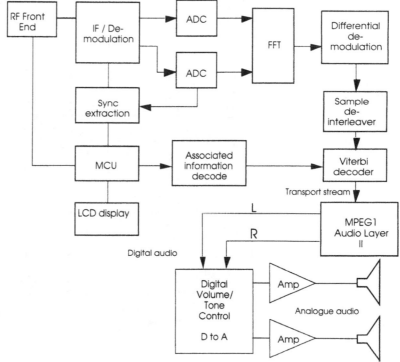

DAB receiver block diagram

14　The multimedia PC

To many, the term 'multimedia PC' is a PC with a SoundBlaster card, a CD-ROM and powerful external speakers that allow multimedia CD-ROMs to be played. Whilst this is undoubtedly a multimedia PC, it is only half the story.

There has already been an upsurge in the marketplace for external modems to add to such PCs to provide a link to the Internet and to allow the user to 'surf the net'. Cards are already available that allow broadcast TV to be played back via a PC screen and thus allow the audio and video to be captured and stored for future use.

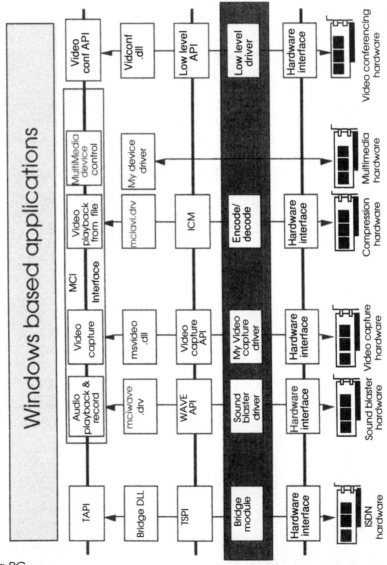

A Windows multimedia PC

If the convergence of the PC, telephone and television set is to happen as predicted, the current architecture of the PC must be expanded to cope. In particular, the integration of computer and telephone, especially with the forthcoming availability of large digital bandwidth straight to the home. Whilst the PC is used to get data from its local disks at a relatively high and uninterrupted rate, it was not initially designed to get data from its communications links, although that is exactly what the multimedia PC of the future will be doing.

This chapter looks at how the IBM compatible PC is evolving and, in particular, how the software environment is changing so it can take advantage of the new communications technology as it becomes available.

The software environment

The previous diagram shows a block diagram of a multimedia PC that can be put together today. It supports ISDN communications, SoundBlaster audio record and playback, video capture and playback and video conferencing. For the sake of simplicity, the obligatory large disk drives, CD-ROMs, and so on, have been left out.

The software environment offered to the applications writer can be split into three: MCI for audio visual control, TAPI for telecommunications and an API for video conferencing. This chapter will go through these environments and look at how they work and interact.

MCI – overview

MCI is the basic mechanism used within Windows to control and access different types of media, such as audio and video. It provides a generalised interface to control devices ranging from audio hardware, video discs and VCR players to movie players. Driver software receives the commands from this interface and performs the various tasks required. For example, via MCI, an application can instruct a video display window to be re-sized. How that is done is of no concern to the application and could be performed in software or hardware. Similarly, an application could instruct a video device to start playing back video. Again it is irrelevant whether the device is a VCR, video or video teleconferencing board — as long as it supports this function. The important point is that the driver software takes responsibility to perform the instructed task in what ever way it wants.

Several command sets are available:

- System commands.
  These are the break, sound and sys info commands.
- Required commands.
  These are recognised by all drivers. They describe the device's capability and status and allow it to be opened and closed.

- Basic commands.
  Supported by all devices, these include basic functions such as load, pause, play, seek and status.
- Animation commands.
  These are essentially video commands with the added capability to stretch and manipulate images.
- CD audio commands.
  Self explanatory.
- MIDI commands.
  These are associated with MIDI devices, such as sequencers, synthesisers, recorders and so on.
- Video disc commands.
  Self explanatory.
- Video overlay commands.
  These control windows re-sizing, video placement, frame freezing and capture. They provide the essential support for video conferencing.
- Waveform audio commands.

The important point to remember is that the MCI interface simply defines what the application requires and not how it is done. It is up to the driver software, in conjunction with the hardware, to determine the best method of implementing and executing the commands.

Consider the video overlay commands. The actual mechanism used to overlay video (i.e. multiplexing or direct access to frame buffers) is not specified and is left to the device driver. The application has no knowledge of how the overlay is performed — that is the responsibility of the driver and the hardware that performs the overlaying.

MCI is a gateway to the many interfaces needed to support multimedia. In its simplest form, as distributed in Windows 3.1, it only supports the control of multimedia devices, such as VCRs, video discs and audio CDs, and support for the sound generation boards, such as Adlib and SoundBlaster. To provide greater integration of video and audio into the PC, Video for Windows is needed. This extends the interfaces supported by MCI to cover video and audio compression and decompression, playback and record from files, and extensive video capture support.

Video control

MCI provides a set of commands that define/control windows and the video insertion or overlay into them.

Defining video windows

Both the window and put commands provide the ability to define and control a video window. This window follows the standard Microsoft window definitions.

Re-sizing and positioning

These are standard window functions. The only difficulty is that the multimedia system needs to know about them so that the video insertion can be re-programmed to work with the window's new location and/or size. This means that either the application or the MCI driver running on the host has to take responsibility for receiving these messages and passing them down. Obtaining the information may require a special graphics module that intercepts and copies all the relevant messages the normal graphics controller would receive.

Multiple screens

MCI can support multiple video windows on the same screen. Additional support for external monitors can be accommodated by the use of a device specific command — MCI_ESCAPE — which programs the multimedia board to direct video to the monitor. This would effectively be a control operation and not involve Windows in any of the video control, unlike the insertion of video into a normal window.

Frame capture

Supported by the save command within the MCI command set for video overlay devices.

Switching sources and destinations

These can be supported by device-specific commands, such as MCI_ESCAPE, to redirect input and output.

There is potential duplication here with the TSPI setphone and setterminal command sets. These commands allow the re-configuration of a telephone device and the input to be switched to a desk microphone or line input, etc. These commands are ideal for video conferencing control but require TSPI to be active. For straight compression/decompression, the MCI_ESCAPE commands that carry out the same function should be used, because there is no guarantee that TSPI will be active, and that a telephony control API be used to redirect a multimedia stream that does not involve telephony.

Video player control

The video player control command set of MCI provides the ability to fast forward, slow motion etc. of video. These commands are not directly supported by MPEG1 and thus require some host tricks to simulate them. These tricks are often implemented by the MCI driver that receives the calls, which can simulate them and use the multimedia supported commands. As such, these commands are a bit of a grey area. If multimedia can support them directly, they should be included. However, currently they cannot and therefore should not be included. One possible option for future expansion is to define them but allow the multimedia system to return a successful code but do nothing.

Video for Windows

Video For Windows (VFW) provides a set of APIs to handle video compression, decompression and manipulation. For the PC, it defines an API called the Installable Compression Manager (ICM) that allows video and/or audio to be compressed or decompressed using either software or hardware. This API allows a host application to access the compression/decompression facilities of the multimedia system.

VFW defines additional video handling functions to complement MCI. These use the MCI command mechanism but extend the supported functions. VFW describes how different video compressors/decompressors can be installed and the corresponding file formats. One interesting point in the documentation is that unsupported file formats can be used and VFW will automatically search for the correct driver to handle that format.

Video compression/decompression — overview

VFW uses ICM, which is used by an application or driver to access the compression or decompression routines. A message is sent to ICM, describing the required function which passes this onto the actual software or hardware that does the work. Access through ICM is assumed to be non real-time; the source and destination are likely to be a file. It also implies that the application is responsible for synchronising audio and video playback if insufficient processing power is available to keep up at the full frame rate.

The next diagram shows how the system routes data to and from the compression driver. Although it states that an application communicates directly, the actual interface is via the MCI interface. Therefore the application is more likely to be a driver that interprets the MCI commands than a true application. The next diagram gives a different view of this mechanism, using interfaces and modules.

An application may call the compression or decompression routines indirectly by sending an MCI command to playback video to a driver, such as MCIDRV. This, in turn, would use the decompression module to take the AVI file and decompress it.

The decompressed and compressed video data is supplied via buffers allocated by the application or driver and it is up to this module — not the decompression or compression modules — to route the data to the video display.

The diagrams show combined compression and decompression modules because they are usually combined. This is not mandatory and separate interfaces and libraries are allowed.

The standard modules supplied by Microsoft to support the MCIAVI driver rely on the driver above ICM to supply and remove the frame buffers. This places the responsibility and loading onto the host processor, rather than move it off board. There are options which allow the decompression modules to directly access the frame buffers and draw directly, rather than pass a completed frame back to the application which would then draw the picture.

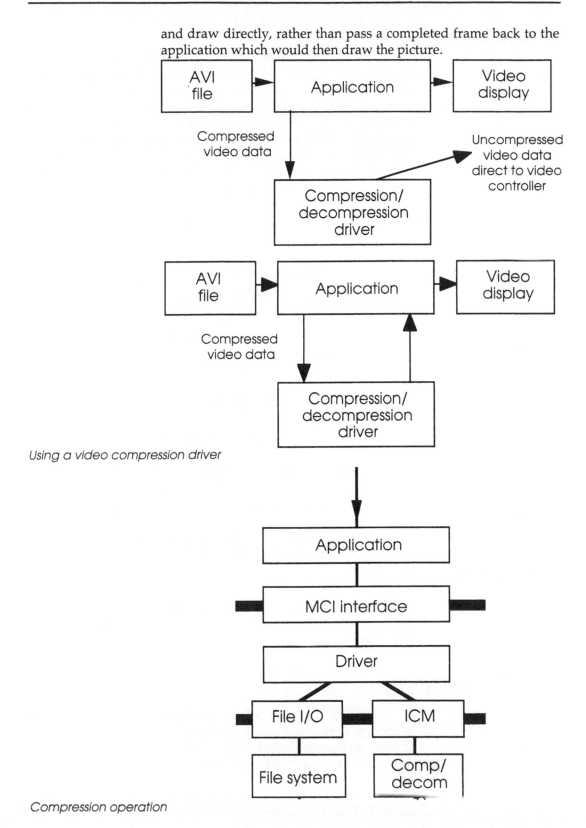

Using a video compression driver

Compression operation

MCIAVI driver

This is a driver supplied by Microsoft which provides basic playback facilities for .AVI files. It allows applications, such as MOVPLAY1, to playback video and audio from an AVI file. In scope, it is the main co-ordinating software that ties in and handles the compression selection and control. The driver is acting as a simple video device which outputs its data into the application window and plays back audio through a SoundBlaster, Adlib, or similar sound card. It is the main driver used to playback .AVI files and many video applications use it in conjunction with video capture drivers etc.

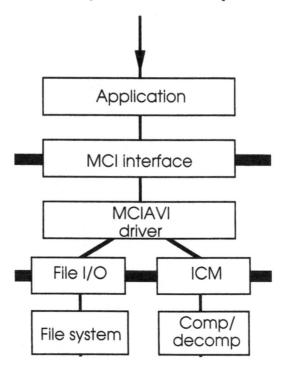

MCIAVI operation

As can be seen from the diagram it is very similar to the previous diagram and accesses its own compression and decompression modules.

TAPI and TSPI

In 1993, Intel and Microsoft released the first draft of the telephony application programming interface (TAPI) and the telephony service providers interface (TSPI) for the Windows environment. These were the first specifications that defined how telephony and PCs should work together. The TAPI document defines how the application interfaces to a telephony service. An interface library called TAPI.DLL links this to the TSPI, which defines how a telecommunications device should appear to the application. In terms of calls and their structure, there is almost a one to one relationship. For a system point of view, TSPI calls are the most interesting because they have to handle and solve the important issues of how the multimedia data is brought into the PC and other problems.

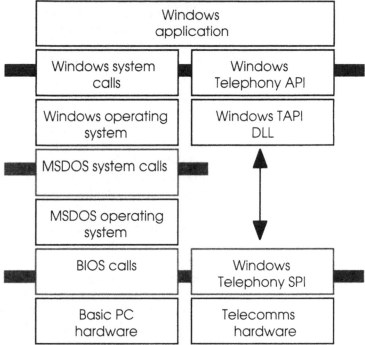

The TAPI/TSPI block diagram

TSPI — overview

There are about 94 TSPI calls within the specification. These are split into three levels: basic, supplementary and extended. Basic services are those required to support a POTS system and essentially allow the PC to make and answer calls. Supplementary services provide additional facilities, such as call transfer, conferencing, call pickup, DTMF tone recognition and other more sophisticated facilities. The extended services are line or device specific and allow additional services to be provided in a regular and organised way.

TSPI call types

The call types are defined as follows:

- Basic services.
  This is a minimal subset of core services. It must be provided by all service providers. The functions contained in basic telephony roughly correspond to that of POTS. The basic telephony subset is listed in the TSPI API specification. Telephone device services are not part of basic telephony.

- Supplementary services.
  This is a collection of all the services defined by the SPI, but not included in the basic telephony subset. It includes all so-called supplementary features found on modern PBXs including hold, transfer, conference, park, etc. All supplementary features are optional. This means that a service provider decides which of these services to provide. The TAPI DLL can query a line or telephone device to discover the set of supplementary services it provides. Note that a single supplementary service may consist of multiple function calls and messages. It is important to point out that the Telephony SPI defines the meaning (i.e. behaviour) for each of these supplementary features. A service provider should only provide a supplementary telephony service if it can implement the exact meaning as defined by the SPI. If not, the feature should be provided as an extended telephony service.

- Extended services.
  These services, also known as Device Specific Services, include all service provider defined extensions to the SPI. A mechanism is defined in the SPI, and reflected in the API, that allows service provider vendors to extend the Telephony SPI using device-specific extensions. Since the SPI only defines the extension mechanism, definition of the extended service behaviour must be completely specified by the service provider.

  The extension mechanism allows service provider vendors to assign new values to enumeration types and bit flags, as well as to add fields to data structures. The interpretation of extensions is keyed from the service provider's manufacturer ID. Special function and call-backs are provided in the SPI that allow an application to directly communicate with a service provider. The parameter profiles are defined by the service provider. Vendors are not required to register in order to be assigned a unique Manufacturer ID. Instead, a utility is provided that allows each vendor to generate a unique vendor ID locally. The unique ID comprises an IP network address, a random number and time of day.

  In addition, a range of values is reserved to accommodate future extensions that are considered common.

Supporting unsupported calls

It is not necessary for a TSPI driver (or service provider, in Microsoft language) to support all calls or even provide null routines for those not supported. The TAPI.DLL provided by Microsoft links the high level TAPI interface with the low level TSPI interface and simulates/emulates any TSPI calls that the driver cannot handle. As many of the functions can be achieved in several ways, the TAPI.DLL may simply use those that the TSPI driver supports. The application would not know about any such substitution. In addition, TAPI.DLL performs parameter checking on the data passed down with each call request.

Asynchronous vs. synchronous completion

There are two types of completion mechanisms for TSPI calls: a synchronous mechanism, where the calling software waits for its request to be completed, and an asynchronous mechanism, where the call is acknowledged immediately but the status information sent asynchronously at a later point in time. It is obviously important that the immediate return for an asynchronous message is not treated as a successful completion and that the message queue is examined to confirm completion before proceeding.

Asynchronous messages

As indicated in the previous paragraph, TSPI calls use two types of communication. The synchronous form is the conventional return message, which indicates that the call has completed, and an asynchronous method, which is used to send asynchronous messages to the upper software layers and for use with asynchronous calls and its success is passed back asynchronously.

Controlling media streams

TSPI supports the control of media streams through the provision of several basic commands but Microsoft does not recommend their use. Instead, the MCI based media stream commands should be used.

The concept of **bearer mode** corresponds to the quality of service requested from the network for establishing a call. It should not be confused with **media mode,** which describes the type of information exchanged over the call of a given bearer mode. As an example, the analogue telephone network (PSTN) only provides 3.1 kHz voice grade quality of service (bearer mode). However, a call with this bearer mode can support a variety of different media modes, such as voice, fax or data modem. Media modes require certain bearer modes. The Telephony SPI only manages the bearer modes by passing the bearer mode parameters on to the network. Media modes are fully managed through the appropriate media mode APIs, although limited support is provided in the Telephony SPI.

One important point: the TSPI driver is responsible for detecting the media mode and activating the appropriate media API to be used for the call and any subsequent changes. This function is performed by the multimedia system and the information passed back to the host as a series of messages.

The bearer mode of a call is specified when the call is set up, or is provided when the call is offered. With line devices able to represent channel pools, it is possible for a service provider to allow calls to be established with a wider bandwidth. The **rate** (or bandwidth) of a call is specified separately from the bearer mode, allowing an application to request arbitrary data rates.

The bearer modes defined in the SPI are:

Voice
: Regular 3.1 kHz analogue voice service. Bit integrity is not assured.

Speech
: G.711 speech transmission on the call.

Multiuse
: As defined by ISDN.

Data
: Unrestricted data transfer. The data rate is specified separately.

Alternate speech and data
: The alternate transfer of speech and unrestricted data on a call (ISDN).

Non call-associated signalling
: This provides a clear signalling path from the application to the service provider.

Although support for changing a call's bearer mode or bandwidth is currently limited in networks, the SPI provides an operation to request a change in a call's bearer mode and/or data rate. The operation is called `TSPI_lineSetCallParams`.

Line definition

A line device can represent a pool of homogeneous resources (i.e. channels) used to establish calls. The Telephony SPI in a client PC typically provides access to one or more line devices.

All line devices support at least the basic telephony services. If the TAPI DLL's client application is interested in using supplementary and/or extended telephony services, the application must first determine the exact capabilities of the line device. Telephony capabilities vary with configuration (e.g. client versus client/server), hardware, service provider software and the telephone network. Applications should therefore make no assumptions as to what telephony capabilities are available beyond just basic telephony services. The TAPI DLL determines the line's device capabilities on behalf of an application via `TSPI_lineGetDevCaps`.

Call definition

Unlike line devices and addresses, calls are dynamic. A call or call appearance represents a connection between two or more addresses. One address is the originating address (the caller), the other is the destination address (the called), which identifies the remote end point or station with which the originator wishes to communicate. At any given time, zero, one, or more calls can exist on a single address. A familiar example where multiple calls exist on single address is call waiting: whilst having a conversation with one party, a subscriber with call waiting is alerted when another party tries to call. The subscriber can flash his telephone to answer the second caller, which automatically places the first party on hold. The user can then toggle between the two parties. In this example, the subscriber has two calls on one address on his line. Since the human user at the telephone handset can only be talking to one remote party at a time in this example, only one call is *active* per line at any point in time; the telephone switch keeps the other call *on hold*. With a line able to model pools of channels, multiple active calls may exist on a call at any one time, depending on the configuration.

The Telephony SPI identifies a specific call by means of a call handle. The TAPI DLL and the service provider both assign call handles as required. Since each needs to allocate a data structure to maintain the state information it associates with a call, each has its own handle. At the start of the lifetime of a call, the TAPI DLL and service provider exchange handles. When the TAPI DLL calls a function to operate on a call, it passes the service provider's handle. Typically the service provider implements its handles as pointers or array indices. This allows simple and efficient access to the associated data structure given the handle. The TAPI DLL uses similar techniques to find its associated data structure when the service provider sends it a message which includes a call handle.

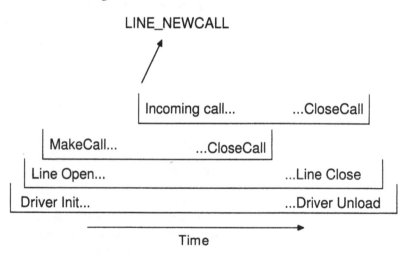

Life cycle of calls

Calls are usually associated with a line, as shown in the diagram. It is essential that the multimedia system maintains this relationship. There are no call specific functions within the TSPI specification. All call related commands are defined as extensions to the line class and therefore do not appear as a separate grouping. In other words, they have been included in the line commands.

Address definition

Each line device is assigned one or more addresses. Think of an address as corresponding to a telephone directory number assigned to the line device. The SPI presents a model where address assignments to line devices are static and will not change until reconfigured at the switch or network.

TSPI_lineGetDevCaps for a line device returns the number of addresses assigned to the line. Individual addresses are assigned address IDs, which are numbers in the range zero to the number of addresses on the line minus one. An address ID is only meaningful in the context of a given line device, so an address is named as the tuple consisting of line and address ID. As described later, a line can be named by its device ID or handle.

The network or switch can configure address-to-line assignments in several ways. Usually, one of the addresses assigned to a line is the line's primary address. A line's primary address is able to uniquely identify the line device, although the address may also be assigned to other lines as a non-primary address. Address configurations recognised by the Telephony SPI include:

- Private.
  The address is assigned to one line device only. An inbound call for this address is only offered to one line device.

In contrast, a bridged address is assigned to more than one line device. Depending on the switch vendor, different terminology may be used such as multiple appearance directory number (MADN), bridged appearance or shared appearance. An incoming call on a bridged address will be offered to all lines associated with the address. Different variations arise from the associated behaviour when using such a bridged address.

- Bridged-Exclusive.
  Connecting one of the bridged lines to a remote party will cause the address to appear 'in use' to all other parties that are part of the bridge.
- Bridged-New.
  Connecting one of the bridged lines to a remote party still allows the other lines to use the address. However, a new call appearance is allocated to the second line.
- Bridged-Shared.
  If one line is connected to a remote party, other bridged lines that use the address will enter into a multi-party conference call on the existing call.
- Monitored.
  The line provides an indication of the busy or idle status of the address, but the line cannot use the address for answering or making calls.

As is the case with line devices, the addresses assigned to a line device may have different capabilities. Switching features and authorisation may be different for different addresses. An application must call TSPI_lineGetAddressCaps to determine the exact capabilities of each address.

Telephone definition

The telephony SPI defines a device that supports the telephone device class as one containing some or all of the following elements, plus additional elements:

- Hookswitch/transducer.
  This is a means for audio input and output. The SPI recognises that a telephone device may have several transducers, which can be activated and deactivated (i.e., taken offhook, placed onhook) under application or manual user control. The Telephony SPI identifies three types of hookswitch devices common to many telephone sets:

- Handset.
  This is the traditional mouth-and-ear piece combination that must be manually lifted from a cradle and pressed against the user's ear.

- Speakerphone.
  This enables the user to conduct calls hands-free. The hookswitch state of a speakerphone can usually be changed both manually and by the SPI. The speakerphone may be internal or external to the telephone device. The speaker part of a speakerphone allows multiple listeners.

- Headset.
  This enables the user to conduct calls hands-free. The hook switch state of a headset can usually be changed both manually and by the SPI. A hookswitch must be offhook to allow audio data to be sent to and/or received by the corresponding transducer.

- Volume control/gain control/mute.
  Each hookswitch device is the pairing of a speaker and a microphone component. The SPI provides for volume control/mute of speaker components and for gain control/mute of microphone components.

- Ringer.
  This is a means for alerting human users, usually a bell of sorts. A telephone device may be able to ring in a variety of modes or patterns.

- Display.
  This is a mechanism for visually presenting messages to the user. A telephone display is characterised by its number of rows and columns.

- Buttons.
  This is an array of buttons. Whenever the user presses a button on the telephone set, the SPI reports that the corresponding button was pressed. Button/Lamp IDs identify a button and lamp pair. Of course, it is possible to have button/lamp pairs with either no button or no lamp. Button/lamp IDs are integer

values that range from 0 to the maximum number of button/ lamps available on the telephone device, minus one. Each button belongs to a button class. Classes include call appearance buttons, feature buttons, keypad buttons, and local buttons.

- Lamps.
  This is an array of lamps (such as LEDs) individually controllable from the SPI. Lamps can be lit in different modes by varying the on and off frequency. The button/lamp ID identifies the lamp.

- Data areas.
  These are memory areas in the telephone device where instruction code or data can be downloaded to and/or uploaded from. The downloaded information would affect the behaviour (or in other words, program) the telephone device.

The telephony SPI allows the TAPI DLL to monitor and/or control elements of the telephone device on behalf of its client applications. Probably the most useful elements for an application to use are the hookswitch devices (e.g. to use the telephone set as an audio I/O device to the PC) with volume control, gain control and mute, the ringer (for alerting the user), the data areas (for programming the telephone) and, perhaps, the display (the PC's display is certainly a lot more capable). The application writer is discouraged from directly controlling or using telephone lamps or telephone buttons, since lamp and button capabilities can vary widely between telephone sets and applications can quickly become tailored to specific telephone sets.

There is no guaranteed core set of services supported by all telephone devices as there is for line devices (i.e. basic telephony services). Therefore before an application can use a telephone device, the application must first determine the exact capabilities of the telephone device. Telephony capability varies with the configuration (e.g. client versus client/server), the telephone hardware and service provider software. Applications should make no assumptions as to what telephony capabilities are supported. The TAPI DLL determines a telephone's device capabilities on behalf of an application via `TSPI_phoneGetDevCaps`.

The next sections describe several scenarios concerned with a host PC using the multimedia system. The examples are based around Video for Windows, although much of the discussion is common to any host implementation. Much of the information is concerned with the various options and product differentiators that an OEM can add to the basic system, as well as descriptions of how the host environment interacts with the multimedia hardware and firmware.

POTS call

This section describes how a simple POTS call can be made by a computer connected to a telephone handset or via a modem to a telephone line. This scenario relies on the telephone device being independent of the computer.

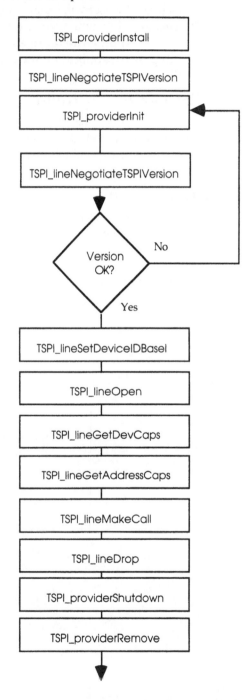

POTS call flow diagram

All that the computer is doing is dialling the number selected from a database, application or control panel. In this case, the call media mode is virtually irrelevant because the computer is not involved in the transmission. The computer may display a 'you are connected. Pick up the telephone and click OK' type of dialogue box but the human then takes control of the call. If the call is to a fax machine, the user will hear the fax noises, make the decision that this is incompatible with a voice call and hang up.

This human involvement is not always the case and, with other examples, the application and multimedia software have to detect the call type and act appropriately.

Installing the TSPI driver software

This series of calls installs the TSPI driver using information from various .INI files and provides the capability negotiation concerning the capabilities expected by the TAPI.DLL and the TSPI driver.

TSPI driver initialisation

This series of calls initialises the driver and ensures that the level of service required by the TAPI.DLL is supported by the associated TSPI drivers. There are two ways of performing this: the first uses the TSPI_lineNegotiateTSPIVersion call, which starts a negotiation dialogue between the TAPI.DLL and the TSPI driver. This mechanism works by comparing version numbers.

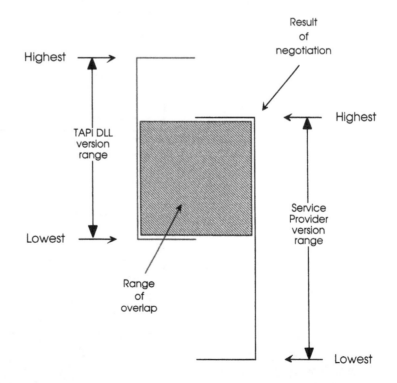

Version negotiation

The TAPI.DLL provides its range of version numbers that it complies with and the TSPI driver returns the highest number within that range that it can support. If there is no overlap, the TSPI driver cannot be initialised and an installation error is returned to the user. After a successful negotiation, `TSPI_providerInit` is performed to actually initialise the driver. Microsoft suggest that a second `TSPI_lineNegotiateTSPIVersion` is called as a final check.

An alternative method is to use `TSPI_lineGetProcTable`, which provides pointers to all the TSPI related procedures the TSPI driver supports. This is slow and, although in the specification, is unlikely to be used.

The `TSPI_lineSetDeviceIDBase` call is used to offset the device ID numbers so that they are the same for both the TAPI and TSPI calls. This is primarily of use with a system that has multiple TSPI service providers. Normally, an application references n lines from 1 to n. Similarly, each TSPI driver does the same and, if there is only one provider servicing all the lines, there is no problem. If there are multiple drivers, their numbering scheme will overlap. This call will assign part of the overall line numbering to each TSPI driver, as needed.

Opening a line

This sequence is straightforward — but there are implications for the application software which interfaces to the TAPI level. The key sequence is that after the line is open, the application checks that both the line and the address are suitable for the type of telephone call to be made. Note that opening a line is not the same as making a call. With this example of a POTS call, it is assumed that all lines can support this level. However, it is up to the application to check before use.

An alternative to checking after opening the line is to use either `TSPI_lineConditionalMediaDetection` or `TSPI_lineSetDefault-MediaDetection` to qualify the line's capabilities before opening. The two approaches would provide information back to the user in slightly different ways. With the capabilities check after the line opening, the user would be informed about the services and media that the line supported. If this did not meet his requirements, the alternatives could be displayed in a dialogue box. With the media detection method, the user would only be told whether a line was found. In the case of failure, the application would need to use the capabilities calls to determine what services the user could utilise.

Making a call

This part is relatively straightforward. It is where the number is dialled and, hopefully, the connection made. Opening a line and making a call are two separate operations that should not be confused. When a line is opened, it does not mean that the call is made. You cannot make a call without opening a line first.

While a call is made, the call status changes as the procedure goes through its various stages: dialling, ringing, connected, disconnected, and so on. This information is asynchronously passed back and, by using `TSPI_GetCallInfo` and `TSPI_GetCallStatus`, information on the status of the call can be obtained.

Status messages

Whilst not immediately part of the flow to make a POTS call, there are two TSPI calls which can set up a message filter to allow only certain status messages to be returned asynchronously and support direct requests for information.

`TSPI_lineGetLineDevStatus`

This call returns the current line status information back to TAPI.DLL and through to the application. This is a synchronous operation which returns a complete definition of the line status. It is used by applications that do not or cannot use the asynchronous messages passed back through the `LINE_LINEDEVSTATE` call back message. Applications can get the line status by using `TSPI_lineGetLineDevStatus`.

`TSPI_lineSetStatusMessages`

This call tells the service provider which status messages should be passed back through TAPI.DLL to the application via the `LINE_LINEDEVSTATE` callback message. It therefore defines the data that is important to the application. Typical status messages are ringing, connected, disconnected, and so on. These messages are asynchronously passed back through TAPI.DLL and it is up to the application to ignore them or use the information to display suitable messages to the user. This call filters the data passed in this way.

Closing a connection

At the end of the call, or in response to an error condition, the line is closed either with `TSPI_lineDrop` or `TSPI_lineClose`.

`TSPI_lineDrop` is used to end the call either at the end of a conversation or transmission or in response to a problem such as the other party disconnecting, wrong or sudden media mode change e.g. voice to fax, or other similar mechanism. An alternative would be to use `TSPI_lineClose`.

Shutting down the TSPI driver

If there is no further use for the TSPI driver, it should be shutdown and removed using either one of the commands `TSPI_providerShutdown` or `TSPI_providerRemove`. This last call removes the TSPI driver and should be executed when all related calls and sessions have completed. It is paired with `TSPI_providerInstall` with calls and sessions in between.

Making a fax call

Making a fax call using basic telephony is a variation on the previous example. The main difference is in the addition and control of a media stream to direct the fax data into the appropriate application.

Most fax modems used in the PC environment are connected either through one of the physical serial ports (COM1 or COM2) or use a virtual device driver that simulates such a port. This method can also be used to provide access for the fax data stream with a call set up and controlled through TAPI.

The changes needed are the inclusion of a virtual device driver, to provide a virtual comms port for the fax data, and the control of this port by the TSPI driver. For example, the TSPI driver initialisation could either install the fax port or check that it is present as an additional part of its initialisation. If the port was not present, this could be reflected in either the results of the initial negotiation or the line and call status information.

The virtual comms port is enabled and disabled by the TSPI driver sending a message to the comms port driver in response to either a connected or a disconnected call status. An example enabling sequence and flow chart are shown.

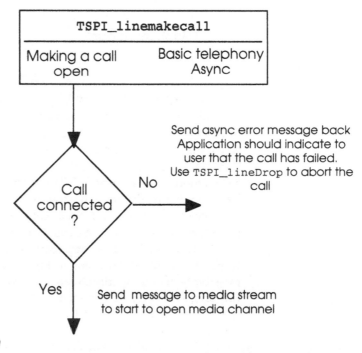

Enabling a media stream

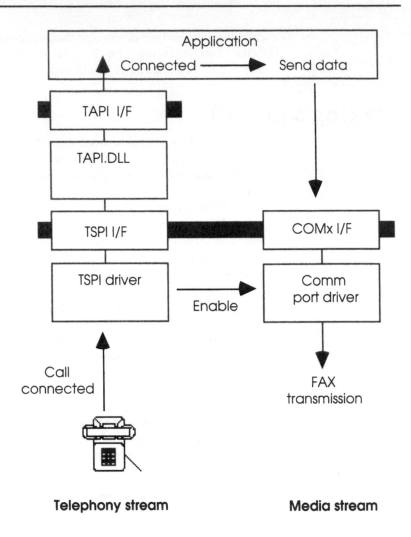

Telephony stream   **Media stream**

Enabling a fax

A further facility must be provided: media mode recognition. The TSPI driver must be able to recognise this and pass the mode back to TAPI.DLL as necessary. With a fax call, this is relatively easy as the called machine will immediately start to transmit and send protocol information. The driver must be able to differentiate between this and a data modem or a voice call. This in itself can be difficult and may have to rely on additional hardware to help. It is hard to determine if a voice on the phone is from a real life person or part of a recorded message from an answer machine or call routing system.

However, before an application can start to use the COMx port to access the fax data, it has to know which port to use. With the possibility of other media streams, such as video, audio and data, it is important that there is a defined method for an application to identify the media stream associated with the incoming call. This process is discussed next.

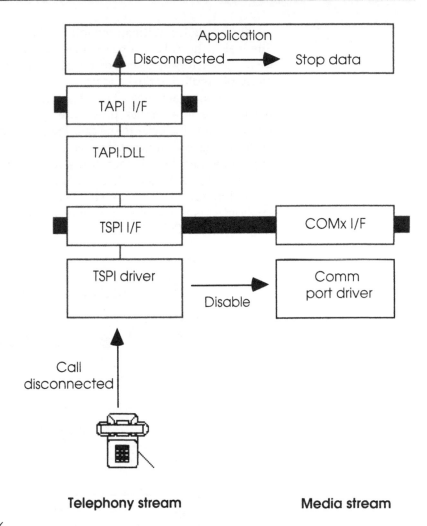

Disconnecting a fax

Identifying the media stream

The media stream can be identified by the application in several ways, ranging from hard coding to use of special TSPI commands to allow the TSPI driver to allocate the media stream.

Each call has an associated media mode which can be used to select which host application is interested in a call — useful if multiple applications are interested in any calls that come on the same line — and to help select which media API will be used to bring the voice, data, and so on into the host environment.

Assisted telephony

With assisted telephony, the host application is not interested in bringing the call information — voice, fax, modem and so on — back into the PC. This information remains external and all that the application does is to use TAPI and the multimedia to act as an intelligent dialler. With an outgoing call, the application opens the line, makes the call and, once connected, takes no further part except for disconnecting at the end of the call — although this could be done by simply replacing the handset.

The diagram shows how this is handled for an incoming call, where there is a similar procedure. The application declares an interest in the line prior to the call. TSPI then receives notification of an incoming call and passes this up to the host application. The application decides whether to answer it or not. If it does decide to answer it, the TAPI_lineAnswer is used. Once the call is picked up, the multimedia system must cope with the media mode because the application is not interested or not capable of accepting the media stream into the PC.

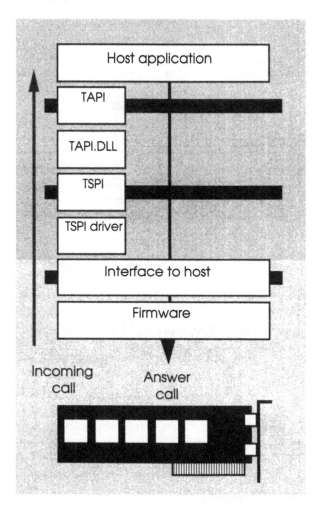

Assisted telephony

This type of application is extremely low level — essentially applets (small applications) that allow external telephone dialling.

Hardwired media streams

A simple but restricted way of allocating media streams to lines and calls is to effectively hardwire the relationships into both the host application and the TSPI driver. In this way, when the host application answers the incoming call, it already knows how to access the media streams. The disadvantage with this approach is that the application is now totally dependent on having the corresponding TSPI and media stream driver support to work correctly.

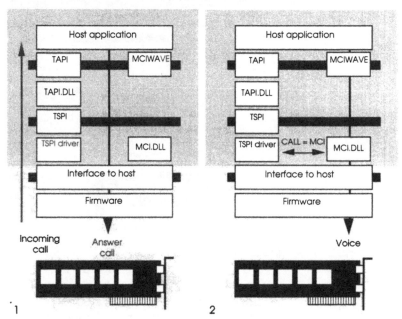

Hardwired media stream for voice only

Configurable media streams

The more normal method within TAPI to allow a host application to identify the associated media stream(s) for a call is to use the TAPI_lineGetID call, which allows the application to ask for the device ID for a given call and media. An example of how this is done is shown both as a diagram and as a flow chart.

The diagram shows the operation as two stages: in the first stage, the incoming call is signalled up to the host application through TSPI and TAPI. The call may have its media stream(s) identified but this is not usually the case. The first decision that the host application has to make is to whether to answer the call or not. In this example, it is assumed that the application does want to answer the call and sends down the TAPI_lineAnswer command.

In the second stage, the host application has to determine both the media mode and the media stream that it should use to access the media stream data. This is done using TAPI_lineGetID. The host application sends this command down through TAPI, specifying the line it is interested in and the media mode. If it receives confirmation that the media mode is supported, it will also be provided with details of the media stream ID that should be used.

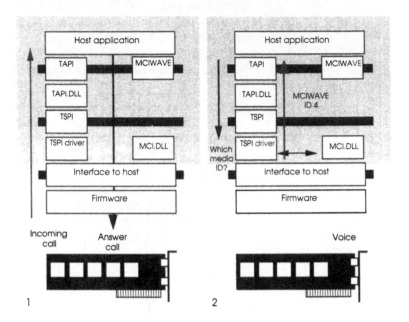

Interrogating TSPI and beyond

If multiple mode streams are supported, the host application may have to iterate the `TAPI_lineGetID` procedure for each media mode. It must also be prepared to do so if the mode changes during the call e.g. the other party sends a fax during a voice call.

The flow chart shows an example flow for this process. Before the call can be answered, the application must be running and declare its interest in line or lines to TAPI. This is essential to allow TAPI to inform the application when an incoming call has arrived.

Switching video sources

The general principle behind switching video sources is the same as for switching audio sources: a button on the telephone or the host GUI generates a message to switch the source which arrives at the host application for processing. This message is then sent to the multimedia system which performs the switch.

The difference with the video scenario concerns which API should be used by the host application to instruct the multimedia system to make the switch. One solution is to use the `TSPI_phone` commands to redirect the video as is used for the audio source switching from handset to speakerphone. The arguments for this approach are:

• The video sources are integral to the video telephone and are thus part of the telephone system.

- Controlling the video source through TAPI makes sense when the video is displayed on an external monitor and not on the PC itself. The external display is treated as part of the telephone device.
- The video window is essentially the output from the video stream. Its contents are determined by the telephone conversation and control commands — these should then be routed through the TAPI/TSPI APIs.

The alternative is to use an MCI_escape command and communicate through the MCI driver used to control the video window on the PC that contains the video data from the H.320 call. The arguments for this approach are:

- This API is used to freeze and capture the video image, so it is the natural candidate for the video source control.
- MCI compatible applications can use/access these functions without having to use or know about TAPI.

In reality, it is likely that both methods will be used by applications. Video conferencing would use the TAPI/TSPI route while multimedia applications doing MPEG encoding, for example, would use the MCI route.

Mixed source dialling

With a multimedia system that has a handset as well as a host application, it is possible (and desirable) for the user to be able to make a call either direct from the application or from the handset — or even alternate between them. This is not as easy as it may seem and many systems do not support it.

It can be achieved by following the normal technique of routing all handset button pushes etc. to the host, rather than the multimedia system itself. The host application collects the data and decides if an external call is to be made and the numbers to use. The application can also receive similar data from the host interface itself and incorporate this data into the stream.

The diagram shows how this works. The user in this example clicks on the screen to simulate the pressing of the number 9 button. The application gets this information and starts to make a call, in the same way that pressing 9 on the handset would start one. The user then presses the 3 button on the handset. This generates a message to the host stating that the 3 button has been selected. The application receives it and, because a call has already started, adds the number 3 to the list of dialable characters. The host application continues the command sequence with the multimedia system to make the call. The multimedia system treats the handset button pushing as an independent activity not related to the call. The host application is responsible for making the connection and redirecting the information back down to the multimedia system.

This can also apply to off- and on-hook signals. The advantage is that the application can decide exactly what replacing the handset actually means, in the context of a call. For example, if a user starts with a voice call and wants to receive a fax from the other party, he should be able to replace the handset without terminating the call — i.e. the on-hook signal must be ignored. By sending it up to the application, the application can decide whether the call should be ended or, as in this case, that replacing the handset is simply a matter of convenience.

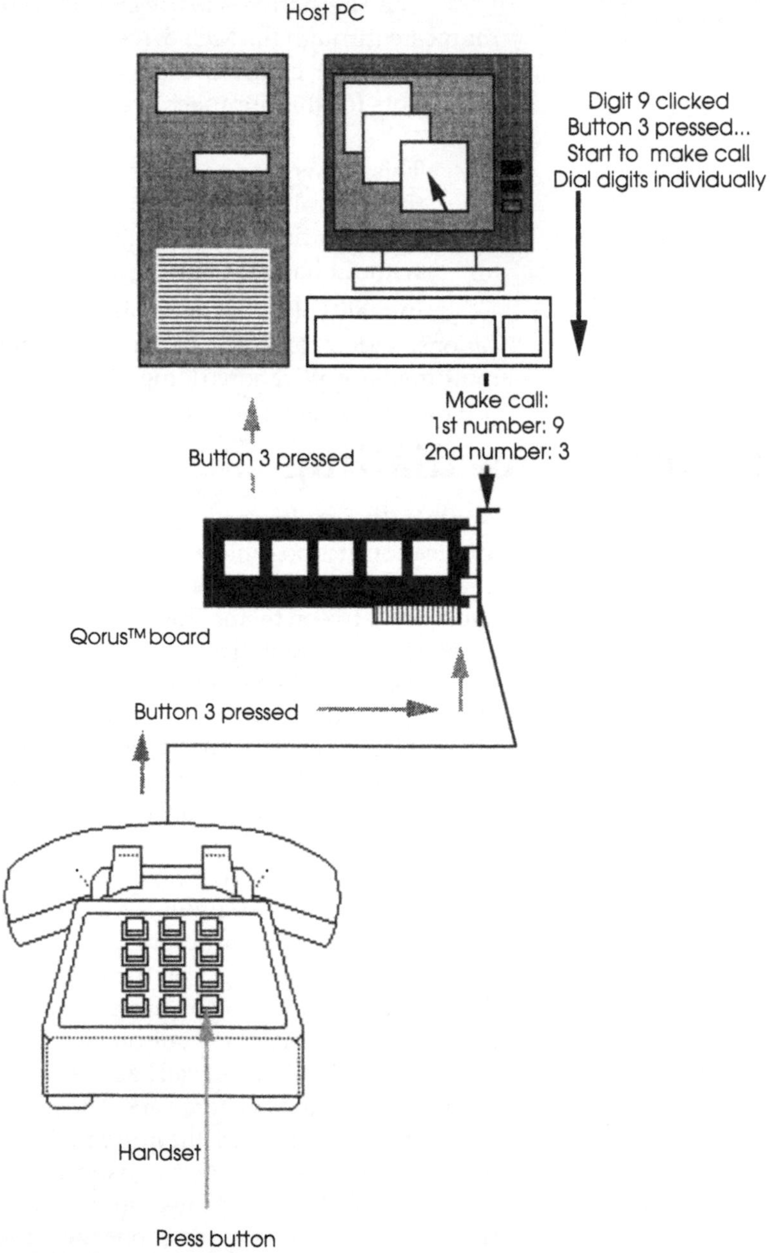

Host PC

Digit 9 clicked
Button 3 pressed...
Start to make call
Dial digits individually

Make call:
1st number: 9
2nd number: 3

Button 3 pressed

Qorus™ board

Button 3 pressed

Handset

Press button

Mixed source dialling

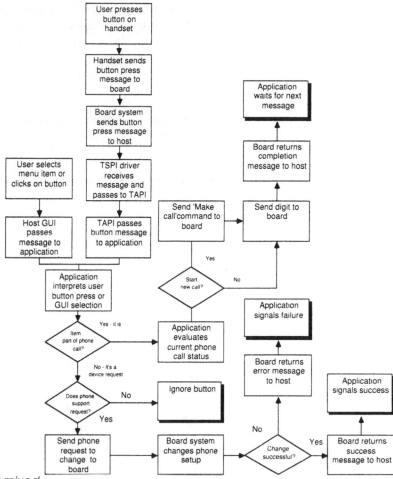

*Example flow chart for mixed
source dialling*

The flow chart gives an overview of the processes involved. It is assumed that the multimedia system is running, a host application is running and has declared an interest in messages from the multimedia telephone devices to the TSPI driver via TAPI. This is necessary to allow the messages to be routed through to the expectant application.

The application has responsibility to order the messages from the host GUI and the handset, and thus construct the resultant telephone numbers and commands from these inputs. As these inputs occur asynchronously and follow very different paths to the application, it is very difficult for an application to correctly sequence digits if the user alternates and selects very quickly. In normal use, this should not occur. Alternating between the host PC and the handset would require some delay and this should be sufficient to allow the application to correctly sort the incoming data. However, it does not prevent a demented user from trying to upset the system this way by simultaneously pressing handset buttons and clicking away on the host. Such situations cannot be catered for 100% of the time.

The chart also shows that each message is effectively treated as an independent entity. The first valid digit received from the handset should be sufficient to start a call if the application wishes to faithfully reproduce the handset's function. Subsequent digits are sent — just as they are when using a handset on its own — down to the multimedia system to continue the call.

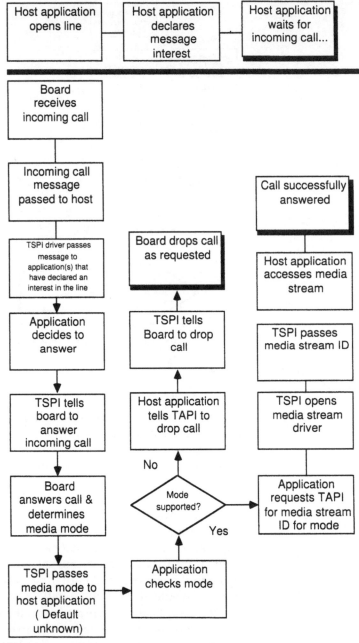

Example flow chart for determining media streams for a single mode.

If the handset is used only as an input device, some other signal must be used to tell the application that it has all the digits and that it should go ahead and make the call. This may be a dial command button on the host GUI or a special handset function.

If an application wants to simulate a handset but the actual telephony device must have a complete number before dialling, software somewhere below the application may have to perform a conversion. For example, the host TSPI driver could start the call if a new digit had not been received after one second. The break would indicate that the number was complete. This would give the illusion of a handset started call, without the need to send the number digit by digit.

Determining media modes for outgoing calls

These examples describe the principles for incoming calls. The same approach should be used for outgoing calls, where the other party may determine the media stream to be used for the conversation. Although the call making application can assume that the other party is an H.320 terminal supporting CIF and G.711 audio, and set up the access to the media streams for this, the other party may be a ordinary handset. It is prudent to use the TAPI_lineGetID call to establish the media streams after the call has been made — and not before.

Data communication

Data communication is complicated because of the different methods that can be used when a data channel is provided within an H.320 call. Within the Windows environment, data communication is performed in one of two ways:

- Using the facilities provided within a network: e.g. file transfer and sharing, compound or shared documents, and so on.
- Using the data channels directly by dedicated applications to transfer files, data, and so on. Such applications include Kermit, Procomm and Timbuktu.

Data communication via a network

There are several ways of transferring data transparently between file systems, PCs and applications. Within the Windows environment, the key mechanism is to implement a local area network which can then be used by protocols such as DDE (dynamic data exchange) and OLE (object linking and embedding) to enable applications such as Excel and Word to embed and link data across the network, and network operating systems such as NetWare, Windows for Workgroups and LAN Manager to perform file transfer and sharing.

By making the data channel available within the ISDN framework appear as part of the LAN, all the protocols and higher level software can be utilised without modification. Fortunately, the interface definitions to link into the appropriate network operating system are available. The most popular are Microsoft's NDIS and Novell's ODI.

NDIS — network communications driver

The NDIS interface definition defines a protocol that allows any Ethernet adapter manufacturer to interface and support Microsoft's LAN Manager network by providing a board driver that understands this interface and protocol:

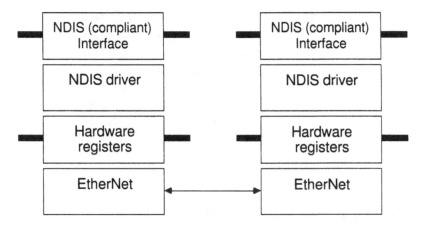

NDIS structure

ODI

ODI is the equivalent for the Novell Netware network:

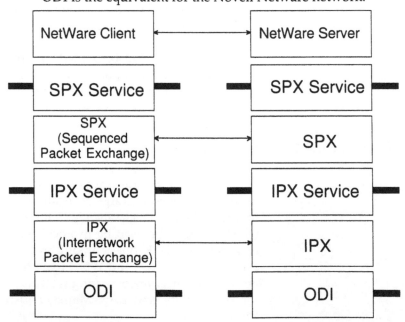

ODI structure

In both cases, it is possible to use these definitions to route LAN traffic onto a different data channel, where all recipients use the same LAN network protocol or can at least understand it.

Restrictions

The successful use of a LAN and technologies such as OLE2 makes several assumptions:

- Each node is linked via a common protocol stack.
- The physical connection and/or transport between nodes is the same. Routers can be used to link different media e.g. Ethernet and ISDN.
- The network can understand and support data exchange technology e.g. OLE2.

Whilst in a homogenous environment this is not a problem — each node uses the same stack, and so on — there are many mixed networks where this is not the case and the advantages offered by the combination of network and data sharing are not obtainable.

Multipoint support

Multipoint support within a LAN is based on the broadcast and sharing capabilities within the LAN network protocol that has been used. This is often achieved through a series of point-to-point communications, which may not be the most efficient way to achieve this type of distribution.

It does not support the protocols needed to control a multipoint control unit (MCU) as is necessary when video conferencing with a large group.

Data communication using dedicated data channels

An alternative method of communication within a PC environment is with dedicated applications and data channels, such as the combination of applications such as Procomm, Kermit, Eurofile with analogue modems and ISDN.

Whilst providing valuable services, this method does have its own restrictions:

- It only supports point-to-point communications.
- Successful communication requires a common communication channel, supported and understood by both parties.
- The applications involved must also understand the low level data transfer and how this information is interpreted. This can often only be done by running the same application at both ends.

Supporting video conferencing

Whilst largely LAN based data communications paths are fine for most PC to PC data communications, they are not sufficient on their own to support other multimedia technologies, such as H.320 video conferencing. Some of the issues that have to be addressed are:

- Provide a standard way of encoding data within an H.320 call. Ethernet has been successful because it defines how information is transferred and removed from the network.
- Provide support to allow LANs to be extended using ISDN, and thus permit the use of LAN based groupware and other data sharing technologies.
- Provide support for efficient multipoint video conferencing.
- Provide interoperability between platforms without depending on common applications.

The T series of specifications provides an open international standard for these requirements and its implementation will form part of the EMI.

T.120 series

The T series is a set of standards that addresses audio graphics conferencing issues:

- It defines the method of encoding data using the MLP protocol within H.320.
- It provides a multipoint communication service for handling and controlling multi-way conferences.
- It defines a generic conference control layer, which includes the conference set-up and tear down, together with maintenance of the MCS layer.
- It provides an API for certain application activities such as binary file transfer and chalkboard.

The motivation behind the T series is to provide consistent interoperability between any terminal capable of supporting audio graphics communication, including an H.320 video telephone, without any participant assuming any prior knowledge of the other system. For example, a PC based video telephone could video and graphically conference with a standalone video terminal. The H.320 protocols would ensure interoperability between them concerning the video and audio, whilst the T series specifications would ensure that both could understand how to create and use a common chalkboard. Neither party would necessarily need to have the same application: the T series ensures interoperability by defining the communication and data transfer, and extends this up to the application layer by defining how mouse movements etc. are transferred.

The standards also cater for network support for fast and efficient distribution of the data stream within a multipoint conference. In this way, an application that uses these standards can interoperate with any other T series cognoscente terminal, irrespective of its configuration or construction.

However, it could be argued that this approach falls short of what is available already within the PC LAN network. The collaborative software far exceeds the chalkboard, file transfer and still image that is supported within the T series. Multipoint communication is again supported — albeit not to the same degree — within the network. If this is the case, why has so much effort been placed in the T series specifications? There are several reasons:

- The PC LAN environment assumes a common network stack and a standard physical communication layer, such as Ethernet. These standards are dominated by Novell and Microsoft with the PC Windows/DOS environment, AppleTalk within the MAC world and NFS within UNIX. None of these are directly compatible except through the use of routers and gateways.

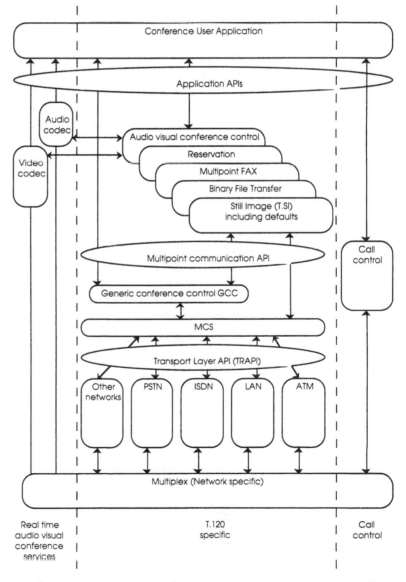

Proposed T.120 API structure

- The multipoint communication is limited to being file based and defined within the network protocol. This is limited to data distribution and does not include, as standard, multipoint control concepts such as a chairman within a multiway video conference. The ability to select which screens are used and which audio is routed is an essential part of a successful video conference system — and this is included within the T series.
- The recognition that video conferencing encompasses more than just PC based systems.

However, the PC LAN environment cannot be ignored. A way forward is to use the lower levels of the T series to define a standard data transfer and encoding mechanism which could be used either as a basis of a PC LAN network link or for a full T series implementation. If the parties can support a network type link, they can negotiate and agree to do so. If this is not possible — the other party may not be a PC, for example — the full T series can be implemented and thus the basic application functionality can be supported.

It is possible that the T series application level API can be supported under existing data sharing technologies, such as OLE and OCE. Currently, Microsoft have committed to supporting and implementing the T series within a future release of Windows. There are several ways that this could be done whilst preserving interoperability with existing software.

Incorporating the T series

The T series of standards contributes several key elements into this framework. The lower levels can be used to define a standard data encoding and transport mechanism which can be understood by any T series compliant terminal. This provides the standard data communication that the lower levels of Ethernet give: it defines the data format and transport. Without this, it becomes virtually impossible to transfer data between terminals and PC based systems.

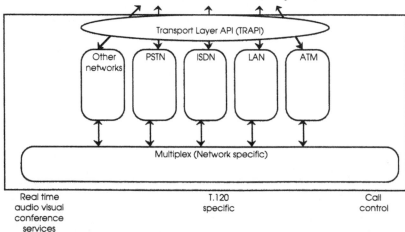

The data transport layers for both PC LAN and T series use

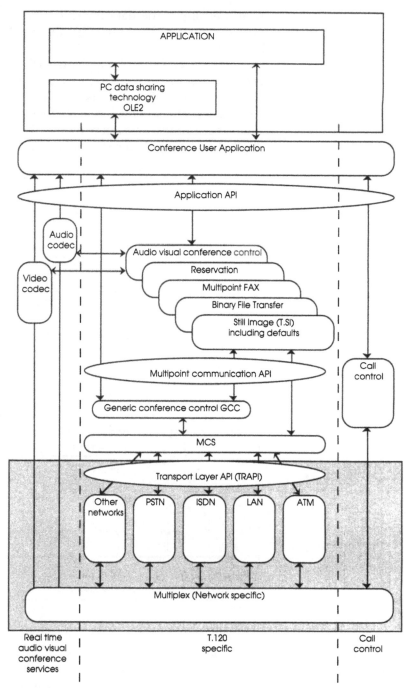

Supporting a full T series stack

With this in place, there are two options for data exchange and collaboration. The first is based around a PC LAN network approach, where the call control is routed through TAPI, the audio and video streams through Video for Windows and the data is routed through an NDIS protocol driver into the LAN Manager network. On top of this sits OLE2, which provides the data exchange and collaboration.

In this example, the data transfer is simply used to replicate an extension of the PC LAN and thus allow all its facilities to be enjoyed by others.

The second example implements the full T series stack and thus provides communication and data exchange between any T series compliant terminal — irrespective of whether it is PC based, a workstation or a simple video telephone. The difference is that the T series application layers can be optionally hidden from the high level application by separating them with the OLE2 layer, for example.

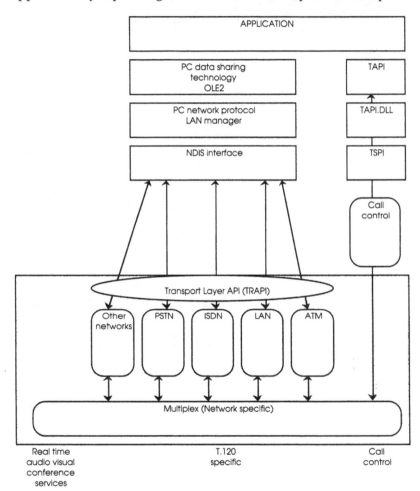

Implementing a PC LAN network using a T series data channel

Index

Symbols

A